Metallic Effluents of Industrial Origin in the Marine Environment

Association Européenne Océanique

A report prepared for the Directorate-General for Industrial and Technological Affairs and for the Environment and Consumer Protection Service of the European Communities by l'Association Europeenne Oceanique

Published by Graham & Trotman Limited
for the Commission of the European Communities

Published in 1977 by
Graham & Trotman Limited
Bond Street House, 14 Clifford Street,
London, W1X 1RD.

for the Commission of the European Communities
Directorate - General Scientific and Technical Information
and Information Management, Luxembourg

Softcover reprint of the hardcover 1st edition 1977

ISBN-13: 978-0-86010-063-8 e-ISBN-13: 978-94-011-7345-2
DOI: 10.1007/978-94-011-7345-2

CONTENTS

FOREWORD

Based on a unique industrial European membership, the European Oceanic Association, **"non-profit Association has as its object the pooling of talents, resources and experience in the marine field with the purpose of promoting in Europe a concentration of efforts to foster the study and exploration of the seas in the light of their industrial exploitation"** *(article 2 of the statutes)*.

The major concern of protecting the marine environment, whilst at the same time taking into consideration the present constraints on economic and industrial development, has led the European Oceanic Association to establish constructive contacts with the Commission of European Communities.

Following an interview in November 1972 with leading representatives of the Commission of European Communities and with the heads of the Environment and Consumer Protection Service, contacts were established which resulted, in May 1973, in the signing of a contract for studies between the Commission of European Communities and the European Oceanic Association.

The essential point of the contract granted to the Association was: **" a contribution to the study of a preventative technology with the aim of arresting the deterioration of the marine environment by limiting and controlling present and future wastes".**

As a result of working groups between interested members of the Association, an area of study was chosen, comprising:

- specific subjects relative to polluting elements or groups of elements, i.e:
  - **chromium,**
  - **cadmium-zinc,**
  - **cyano-metallic compounds;**
- more general studies concerning monitoring systems;
- specific and general recommendations.

The present report gives the essential points of the results of the studies which were carried out in accordance with the terms of the contract, studies which were completed thanks to the appreciable contribution of the industrial members concerned of the European Oceanic Association.

CHAPTER 1

PRINCIPAL REASONS FOR THE PRESENT STUDY AND CHOICE OF ITS OBJECTIVES

1.1. THE PROBLEM OF INDUSTRIAL METAL WASTE DISPOSAL INTO THE MARINE ENVIRONMENT

Because of its extent and volume (1,370 x $10^6 km^3$), the hydrosphere could be considered as an ideal disposal site for most wastes produced by man. This concept, long accepted on a world wide scale, and especially by developed countries, has led today to a grave state of ecological poisoning, aggravated by the long term persistence in the marine environment of certain pollutants, and by the bio-concentration of a whole range of toxic chemicals.

If the marine environment, toxicologically speaking, includes a number of "natural" biotoxins, there are now increasing quantities of toxic substances resulting from human activities being introduced as well.

It appears today, however, in a manner more and more in evidence, that certain pollutants, often industrial in origin, can reveal or accentuate certain bio-toxic cycles, causing disruption of the biomass equilibrium (notably planktonic), resulting in the growth of populations of dangerous marine organisms.

At the human level, there are phenomena, whose origins are now well known, such as "Minamata" and "Itaï-Itaï" diseases, as well as those of more obscure origin, for example, "Ciguatera" disease which is found in certain tropical areas, and all of which are caused by the toxicity of certain pollutants. This is due to the ease by which these pollutants pass along food chains of various trophic levels, causing changes, both biological and structural, to cellular material.

It is thus clear that pollutants disposed of in the sea or carried there by such pathways as streams and rivers, are not only diluted and dispersed by currents, tides, swell and wind, etc., but also become an integral part of biological and biochemical processes and, far from being further diluted, are re-concentrated and propagated with all their original toxic potential.

These phenomena can produce, moreover, a sudden reactivation of some pollutants, caused by re-suspension due to dragging operations, for example. This can occur even with pollutants which have apparently been strongly fixed, for long periods of time, in marine sediments.

If it is agreed that a grave danger exists for the marine environment and its biodynamic potential, then the question remains as to the nature and real extent of the effects of pollution on the biosphere.

It should be remembered that the knowledge of this subject is incomplete, dispersed in many disciplines which are difficult to correlate and often resulting in contradictory interpretations of the data.

However, due to the slowness of reaction of the marine environment and the uncertainty of the fate of many polluting compounds, it is very difficult to determine the real threshold risks and one is left, often, with the feeling that calculated limits have, in certain cases, already been exceeded.

In such a context, without waiting for the necessary fundamental and multidisciplinary scientific research, the delay almost always being incompatible with the need to take urgent measures, the essential idea of a preventative technology appears as one of the only openings possible in finding a solution to one of the most important questions with which the modern world is confronted.

1.2. DEFINITION OF THE TERMS OF REFERENCE OF THE STUDY

The terms of reference of the study have been defined as follows:

- specific subjects relating to certain polluting elements or families of elements;

- more general subjects having a future perspective particularly in measures and monitoring.

a. Specific subjects

Among the metals or metal compounds that could affect the marine environment, the study will deal with the following:

. **chromium**
. **cadmium and zinc**
. **cyano-metallic compounds**

. Concerning **chromium**, the accent will be laid principally on those compounds whose toxicity and stability cause them to be pollutants in the marine environment and which will have to be controlled. Necessary preventative technology will be required.

. **Cadmium** - This heavy metal is on the 'black list' of potential pollutants and is under serious study in many parts of the world. Nevertheless, many important aspects of its interaction in seawater in the presence of **zinc**, appears to exhibit certain synergetic effects (potential or attenuating) of which the repercussions on the environment are far from being understood at the present time.

. The fate of **cyano-metallic compounds** and their degradation products, as well as **metallic cyanides** remains an important problem, even though these latter substances have a relatively short life in the aquatic environment. Because of this, even if the use of these products in the future does not increase at the same rate as at the present, the studies undertaken by certain industrial firms, and particularly by members of the A.E.O., with a view to controlling, at source, effluents containing cyanide compounds, have to be developed in more detail.

The reasons for the choice of these specific subjects for the study will be discussed in more depth in 1.3.

b. **General scope**

The study itself on the elements mentioned briefly above, is completed by various general considerations concerning the orientation of future studies of which the importance and interest to Europe appears incontestable.

. The first point will concern the question, of primary importance, of **measures and monitoring** of metallic waste products.

In effect, in order to produce realistic and efficient measures for the environment, a vast amount of data on the concentration of pollutants, both in the environment as well as in waste effluents, is needed.

At the present time, several classical methods of chemical analysis (principally manual) are known. However, the major weakness of these methods lies in the low number of samples that can be treated and in the delay between sampling and obtaining the results. These handicaps make these methods inadequate for the detection and control of pollution and for giving a rapid warning.

For these reasons, the design and construction of automatic stations for sampling and analysis have become necessary; stations that will transmit the results of the analysis to a central control, either by telephone or telex. In such automatic stations, a number of technical considerations must be incorporated, these being the production of rapid and precise results, and the functioning of the station over long periods of time without surveillance.

In the aquatic medium, especially in sea water where the concentrations of most of the toxic metals lie between 0.1 - 10 μg/l, the sampling method, treatment and measurement must receive careful attention in order to be certain of the accuracy of the results. Certain problems are caused by the fact that some chemical species of the metals concerned are not in an ionic form; also, some may be adsorbed by solid particles in suspension.

The classic methods of chemical analysis (solubilization and extraction) are inadequate, and new techniques need to be invented. Finally, the optimum frequency of sampling still needs to be determined.

. Finally, based on a foreseeable programme of planned construction of **sewage purification stations** in coastal and inland areas, from now until 1980-1985, the question of the efficiency of retention of polymetallic compounds by the different types of stations will be examined.

In effect, in the next four or five years, industry will be obliged to treat the greater part of its own pollutants by itself, though smaller firms will be able to use local purification plants. In this context, it appears necessary, if not in fact indispensable, to know accurately the effect of the metallic waste on the function of these purification plants, both upstream and downstream. At the same time, it will be necessary to know the retention efficiency of the plants for these compounds.

. The problem of **restoring** gravely **polluted areas,** especially those polluted by heavy metals, such as lagoons or other regions where there is accumulation of sediments, will be resolved as much from an evaluation of the degree of toxicity of the sediments with regard to its benthic populations, as from conditions of restoration and repopulation.

It would be interesting to incorporate in this research some fundamental studies concerning the neutralization of toxic compounds by mineral or partially biologic adsorbtion, and their elimination from the sea layer in equilibrium with the atmosphere by sedimentation in the deep sea layers.

1.3 CRITERIA USED FOR THE SELECTION OF METALS AND METAL COMPLEXES STUDIED

Three main criteria are employed in choosing the metals or metal complexes studied in the report:

- the nature and strength of the aggression, considered essentially from a toxicological point of view;

- the degree and forms of occurrence of these elements;

- the future industrial trends in the use of these elements or complexes, up to the year 1980.

The criteria will be specified for each of the pollutants selected for the study.

a. Nature and intensity of the problem

a.1 Chromium

From a toxicological point of view, soluble hexavalent salts of chromium are undeniably the most dangerous, even though recent studies have revealed the importance of the trivalent form (active species in transfer processes). The pathological mechanism is essentially one of ulceration, by direct attack or by progressive fixation of certain tissues (such as mucus membranes and cartilage). Although it is not well understood, the major risk may well lie in the potential carcinogenic effect of certain chromium-based pollutants.

In relation to the aquatic biota, especially that occurring in the marine environment, this particular aspect appears to be the most important, although ulceration can cause a degradation of biota by either or both of two routes:

- a diminution of resistance to infection;
- an influence on the process of fecundity and growth.

For these reasons, chromium and certain of its compounds deserve special attention, even if the nature of the degradation, when they are transferred by ingestion, is not clearly settled at the present time.

a.2 Cadmium/Zinc

Cadmium is always present in metallurgical processes of zinc and is undeniably a health risk, being toxic either by inhalation or by ingestion, the risk proving essentially to be one of exposure for long periods of time to low concentrations of the element.

In liquid medium, contamination of man occurs through contaminated food; pathologically, this is observed in kidney and bone diseases (Itaï-Itaï disease).

A strong metabolic relation exists between cadmium and zinc and the pathogenic risk lies in the accumulation of these two elements.

There are very few studies on the distribution of cadmium in marine biota; however, some effects are foreseeable:

- the known ability of cadmium to be concentrated by living matter will cause a reinforcement of its toxic nature;
- its function on cellular metabolic processes may be too great for the ability of the species to adapt and alter its protein synthesis;
- the resurgence of poisoning from fish underlines the importance of the phenomenon of sedimentation and,

in consequence, the necessity of very strict controls, at source, of effluents containing cadmium.

Concerning zinc, the knowledge of industrial toxicology is less dependent on the risks of poisoning by zinc itself than on the effects caused by the presence of other associated metals, such as arsenic, cadmium and lead.

From the standpoint of physio-pathology, zinc combines with two types of proteins:

. those of certain metallic enzymes,
. those of proteins that carry zinc.

For man and the higher mammals, zinc is in general absorbed little and is mainly excreted through the intestinal tract.

The action of zinc on marine biota is not well understood and interpretation of data is often difficult because of its rapid incorporation. Nevertheless, it is important to underline its potential for fixation by association with other metallic elements, notably cadmium.

a.3 Cyano-metallic compounds

This group includes both simple salts and halogenated derivatives. The nature of the attack is essentially acute and intense in form, and is by direct action at the enzyme metabolic level (formation of complexes of the CN^- group with enzymes). The toxicity depends upon the concentration of the CN^- group. At sublethal concentrations, the CN^- group is liberated and converted into a non-toxic compound, which is excreted by the kidneys.

In the marine biosphere, the cyano-metallic compounds represent a risk due more to the localization of the effect than by widespread dissemination. However, an important problem that remains is to determine at what level of evolution occurs the ability of an organism to enzymatically catalyse cyano-complexes to render them non-toxic.

In the above three cases, the toxic impact on biota will emphasize the modifications of the interaction of various species (chemical intermediates), the metabolic stress undergone and its consequences.

b. Occurence, degree and forms

b.1 Chromium

The salts of chromium are numerous; the element itself can have several valence states, being in either cationic or anionic form. However, there is a possibility of complex associations.

. Chemically speaking, the most common compounds are:

1) **Alcaline chromates and bichromates** -
Being in the form of very soluble crystals, sodium bichromate ($Na_2Cr_2O_7$, $2H_2O$), is the most important. It is used as raw material for the production of other chromium salts, as well as in tanneries and electroplating processes, etc.

2) **Chromium sulfate** ($Cr_2(SO_4)_3$) -
is also often used (in various basicities) by industry, e.g. tanneries.

3) **Chromic acid** -
This is a very powerful acid and has many industrial uses.

. **In metallurgy,** the most important compounds are the **ferro-chromates**, used in the production of special steels; the **metal chromium** is also used in the production of alloys, containing large amounts of it.

. **The production** of chromium and its compounds is obtained from "chromite" ores (composed of about 50% Cr_2O_3), from spinels (Cr_2O_3, Al_2O_3-MgO, FeO) containing traces of vanadium, titanium, etc.

From these various operations, as can be imagined from the above discussion, harmful materials are formed, such as:

- solids, such as ashes, slag, dust, etc.;

- liquids, such as effluents of various dilutions in which chromium may appear in the hexavalent or trivalent form;

- aerosols, muds (slime) etc., which are generally in one of the two principal forms mentioned above.

. **In the production** of chromium and its compounds, **losses** range in the order of 10% of the total quantity available, a small amount of these losses being in the hexavalent state.

. **In the use** of chromium, **losses** range in the order of 25%, as in tannage, for example, but increase to more than 50% in certain chromium plating processes. The chemical form of chromium lost in this way is either in a tri- or hexavalent form. The greater part of these losses can, however, be recovered.

Because of the great variety in stability and solubility of chromium salts, as well as their ability to change from one form to another, it may well be considered that, if the harmful products are numerous, the preventative techniques can benefit by these characteristics.

b.2 Cadmium/Zinc

Many industries produce wastes containing cadmium which are likely to be released into the environment. There is much information on the sources of pollution in available literature. To the list of industries most commonly cited (metallurgical processing of zinc, carbochemistry, production of pigments for paints and enamels, cadmium plating processes, electricity and battery production, photographic and pharmaceutical industries as well as alloy production), industries using vinylchloride for plasticizers/stabilizing substances containing barium/cadmium have to be added.

As has been mentioned above (1.3.a.2), cadmium presents a real risk for the marine environment, as much from a toxicological standpoint as from one of human health.

Because of this, cadmium has already been the object of several penetrating studies. A colloquium held by the Commission of European Communities, in Luxembourg, 3-5 July, 1973, entitled: *"European Colloquium on the problems posed by the contamination of Man and his environment by mercury and cadmium"* (58)*, brought together many scientists. However, there is still a lack of knowledge concerning certain aspects of the behaviour of cadmium in liquid mediums, this point being mentioned by several authors.

One may add to these remarks some statements made elsewhere (cf. l'*Inventaire du degré de pollution des eaux superficielles, rivières et canaux,* Paris, la Documentation Française, 1973), concerning the distribution of cadmium (one quarter of samples taken have a concentration of cadmium higher than the limit recommended by the World Health Organization). **It thus appears necessary to understand more completely the solubilization (microsolubilities) of many of the cadmium salts in waters,** with a view to reinforcing the concept of preventive technologies, especially for very low levels of concentration.

Following a technical exchange of views, it was decided not only to include cadmium in the present study, but also to discuss zinc, and especially the cadmium/zinc relationship.

Indeed, it is already well known that the only sources of cadmium are zinc ores; moreover, the harmful products of industrial uses are those of both zinc and cadmium.

Finally, it is hoped that one will be able, with the aid of radiochemical techniques which are very useful, to study the transfer and localization of traces; to 'follow' these elements, incorporated or not, with relative ease.

* *Cf. Bibliography at the end of volume*

b.3 Cyano-metallic compounds

Whether these elements are in the form of simple salts, halogenated derivatives, or organic complexes, the number of combinations with CN^- is high and they are often found as waste products in effluents.

Many industries are involved in the use and waste production of cyano-metallic compounds, such as:

- metallurgical industries (scrubbing of exhaust gases from blast-furnaces, steel tempering processes, etc.);
- coke and gas production;
- electroplating industries;
- chemical and petro-chemical industries (organic and inorganic dyes, synthetic fibres, etc.).

The cyano-metallic compounds which are the most representative and most used today, are the cyanides of sodium, potassium, copper, zinc, nickel, cadmium, silver, gold and iron. Their physio-chemical characteristics emphasize their mobility and chemical replacement ability.

In an aquatic medium, the cyano-metallic compounds can play a potential polluting role, by association with other polluting metallic ions.

Finally, cyano-organic compounds are present in waste waters resulting from processes which form cyano-nitroxyl compounds (used in synthetic fibres formation), and in wastes originating from the synthesis of certain organic compounds (acrylonitroxyls, acroleins, etc.).

c. Estimated future industrial uses of the metals and metal complexes up to the year 1980

c.1 Chromium

Many reports have been published concerning the quantities of raw material, the production and industrial uses of this element and its compounds. We recall here, particularly, the reports prepared for the Commission of European Communities, by the Kienbaum group in Brussels in 1972, *"Rapport au sujet d'une étude sur les sources et les volumes de résidus et déchets solides, liquides et gaseux de métaux lourds rejetés dans le milieu ambiant, en République d'Allemagne Fédérale et en France, réalisé pour le compte de la Commission des Communautés Européennes"*, and by the Central Laboratory TNO of Delft, by W.H.J.M. Wientjens, *"Le mercure, le cadmium et le chrome aux Pays-Bas: rapport final"*. These studies were to be completed, particularly in France and Great Britain.

In 1970, approximately 900,000 tons of chromite mineral were used in the Republic of West Germany and in France, representing a potential of 450,000 tons of Cr^2O^3 and 300,000 tons of the element (chromium). Taking into account that more than half of this chromium was for industrial steel production, there remained a little over 100,000 tons used, principally, in the production of bichromates. When one considers the small efficiency of the use of chromium and the complexity of production, this last figure is far from being negligible from the point of view of the environment.

Thus, in this context, taking into account the competence and experience of certain members of the A.E.O. in this particular branch of industry, it appeared that it would be extremely valuable to make chromium the subject of preliminary proposals initiating investigations into possible preventive technology.

Finally, it is not possible to foresee, in future trends of the use of chromium, a diminution of industrial uses but, rather, it can be reasonably expected that there will be an increased requirement in Europe: a doubling every 10 years.

c.2 Cadmium/Zinc

In the future, the use and consequently the production of zinc and its compounds will increase greatly. Its principal by-product, cadmium (the production of 10 tons of zinc gives rise to about 25-30 kg of cadmium), can only follow this tendency, making even more urgent the necessity to formulate proper controls for the disposal of wastes containing this element into the marine environment.

The consumption of cadmium in the United States of America in 1972 (8.3 million pounds) was 20% above the consumption for the two preceding years. This increase seems to be reflected in all countries though the level of production, directly correlated with the production of zinc, needs to be interpreted with care.

c.3 Cyano-metallic compounds

A yearly increase of 3% in the production of these compounds can be envisaged from the present moment until 1980, except if the substitution of cyanides by organic compounds increases.

It is also feasible that the future will reveal important changes in the different industrial uses of cyanides, by the acquisition of entirely new techniques.

1.4 SOME COMMENTS ON THE REASONS FOR EXCLUDING CERTAIN METALS

At the present state of knowledge of marine pollution, the heavy metals and their compounds most often listed as pollutants are, in general, as follows;

- mercury
- lead
- cadmium

. chromium
. cyano-metallic compounds
. organo-phosphorated compounds
. arsenic
. fluorine compounds
. silicon

Taking into account the elements or families of elements studied in the present report, the reasons for the exclusion, or temporary non-inclusion, of certain metals and toxic compounds are explained below.

a. Mercury

Considered as one of the most dangerous pollutants to man, mercury is of great interest. However, the knowledge of this danger is well understood, as well for coastal regions as for the total inventory of the sources of mercury pollution.

The Organization for Economic Co-operation and Development (OECD), published in 1974 the results of a study made between mid-1971 and mid-1973 concerning the use of mercury, its emission, biological effects and control (59), as well as a line of political propositions and recommendations to the member countries with the aim of decreasing the disposal of mercury waste. An information interchange between the different member countries was established.

Moreover, a good number of the problems relating to mercury appear to be under control. This is illustrated by the careful studies on the elimination of mercury from exhaust gases resulting from electrolytic processes. There are two distinct regions of application:

. concentrations of mercury less than $5.10^{-3} g/m^3$

. concentrations of mercury less than $1.10^{-6} g/m^3$

A further example is the replacement of certain mercury compounds (alkyl or aryl mercury compounds in particular), by others which are less toxic, in such diverse activities as agriculture and paper manufacturing.

Elsewhere, the production of certain synthetic organic materials (based on acetylene) where mercury was used as a catalyst and which used to get into waste effluents in a particularly dangerous form, has been stopped or controlled.

Methods now exist for rapid detection and precise analytical determination of this element, though at present no automatic method for measuring total mercury is in use. These analytical methods have largely contributed to diminishing the difficulties encountered in the studies undertaken to identify the mechanisms of the transformation of the mineral form of mercury into organic form, the pathways of mercury transfer to biota, and the bioaccumulation by living organisms.

Finally, concerning the determination of thresholds and limiting concentrations for normal needs, a great deal of work has already been accomplished or is in the course of being carried out.

Thus, it seems reasonable (the field of future research being limited*) not to study mercury in the present report; an increase in the potential danger of this element in the near future, except by a natural catastrophe, appears to be improbable.

b. **Lead**

For many reasons, lead is an element of great interest.

- Because of its chronic toxicity, "saturnism", like that of mercury, "hydragyrism", this element has been made the object of many studies.

  Lead is regarded mainly as an atmospheric pollutant and preventive technology is mainly concerned with lead contamination of effluent gases; this has been and is still the subject of detailed studies.

  These programmes are concerned as much with the problem of city environment pollution by petrol engines, as with factories using or producing products based on this element.

  Thus, for example, the production of lead tetraethyl (the first factories were built nearly 40 years ago), and that of lead tetramethyl (produced for at least 15 years), have been subjected to very strict internal control, of an exemplary nature.

  Parallel to this, great precautions have been taken in order to avoid waste releases of all forms of lead, both in gaseous and liquid effluents. The use of lead compounds in petrol engines has resulted in an enormous number of fundamental and technical investigations.

- The *Symposium on the health problems caused by lead present in the environment,* organized by the Commission of European Communities in Amsterdam, October 2-6, 1972, was concerned with the detection and monitoring of lead and its compounds. It emphasized the considerable work that has been already carried out on this subject in the analysis of trace quantities of lead in blood, air, drinking water, etc. A consensus was even reached as to the methods of sampling and measuring.

  For the marine environment, a method which appears very useful for the detection and measurement of lead is that of isotopic dilution, which can be used for concentrations ranging in the order of 50 ppb. Much

* *Cf. European Colloquium on the problems posed by the contamination of Man and his environment by mercury and cadmium. Luxembourg - July 1973.*

work is being carried out at present on the prevention of sample contamination by external sources of lead; this is especially necessary when measuring very low concentrations of this element.

It must be remembered that lead (Pb^{++}) precipitates at pH 6, so that lead salts are generally rendered insoluble in seawater at pH 8.1 ± 0.2, which is in contrast to their reactions in fresh water. Several articles on marine pollution by lead have established that lead in waste effluents rapidly sediments in the sea; it can probably be said safely that, with the possible exception of bottom living fauna, there is only slight probability of contamination of pelagic species.

The points briefly outlined above constitute the principal reason for the non-inclusion of this element in the present study.

c. Arsenic

From chemical and toxicological standpoints, arsenic is of real interest, and if this element has not been chosen, it is rather because of a lack of immediate opportunities than because of a lack of interest.

It must be remembered that arsenic compounds have been nearly completely abandoned and have been replaced by organic compounds, for example, as in the case of agriculture. The harmful effects caused by these replacement products have not always been immediately visible and whether or not they will result in unforeseen reactions, it is too early to say with any degree of certainty at the present time.

However, the problem caused by the use of by-products (residues from other productions) having an arsenic base, resulting from the treatment of certain ores, is one of a general nature. The same problem will appear in the use of products recovered by installations using preventive techniques in the future.

d. Fluorine compounds

The problem of residue fluorine compounds has been well studied to date, chiefly in relation to the aluminium and fertilizer production industries, as much because of the relative scarceness of the "spath-fluor", as in order to diminish the harmful effects of atmospheric dust produced by this element. It should be noted that a decrease in atmospheric dust will cause pollution of aquatic mediums to increase, but in a less dangerous and more easily treatable form.

In the near future, when there is a general capability of controlling pollution at its source, the problem of prevention of pollution by fluorine and its compounds will be a part of the general problem of improvement and recycling of materials.

e. Silicon

Silicon and its compounds represent a serious problem but, as with arsenic, it will possibly be treated at a later date.

CHAPTER 2

ORIGINS, FORMS AND CHARACTERISTICS OF CHROMIUM POLLUTION

CHAPTER 2 - ORIGINS, FORMS AND CHARACTERISTICS OF CHROMIUM POLLUTION

2.1 SOURCES AND PRODUCTION

a. Types of minerals

Essentially chromium is obtained from chromite, a "spinelle" type of mineral (iron and magnesuim chromo-aluminate, (Fe, Cr, Al)$_2$O$_3$. (Mg, Fe)O).

Three main forms of this mineral are known as:

- **Metallic Chromites** containing about 48% Cr_2O_3.

- **Chromites** used in industrial **refractory** processes, containing about 36% Cr_2O_3.

- So called **chemical chromites** having a Cr_2O_3 content of approximately 45%.

It must be noted, however, that the average concentrations mentioned above are far from being precise. For chemical uses, it is quite possible to use so called "rich" (as are certain Russian chromites) or "poor" chromites (as are certain minerals from the Transvaal).

Metallic chromites can have a **chromium iron ratio** as high as 3/1 or as low as 2/1.

b. Production of metal and metal compounds

The large range of metal and metal compounds of chromium used in industry may be separated into two main groups:

- Production of **ferro-chromes** and metal compounds for the metallurgical industry, chromium metal can also be included in this group. (Chromium metal is sometimes included in the production statistics for chemical compounds).

- Production of **chromium salts** and intermediate products such as chromate, dichomate, chromic acid, chromium oxide and chromium sulphate.

To these two main groups of production, must be added that of **refractory** materials which are reduced in the major processes of mixing and cooking.

Figure 2.1.b is a schematic diagram of the main supply routes and treatment processes in the chromium industry.

In **annex 1** at the end of this report, diagrams A.1-2a, A.1-2b1, 2b2, 2b3 and 2b4 show some simplified models

FIGURE 2.1.b - SOURCE AND TREATMENT CHANNEL FOR THE CHROME INDUSTRY *(Private communication)*

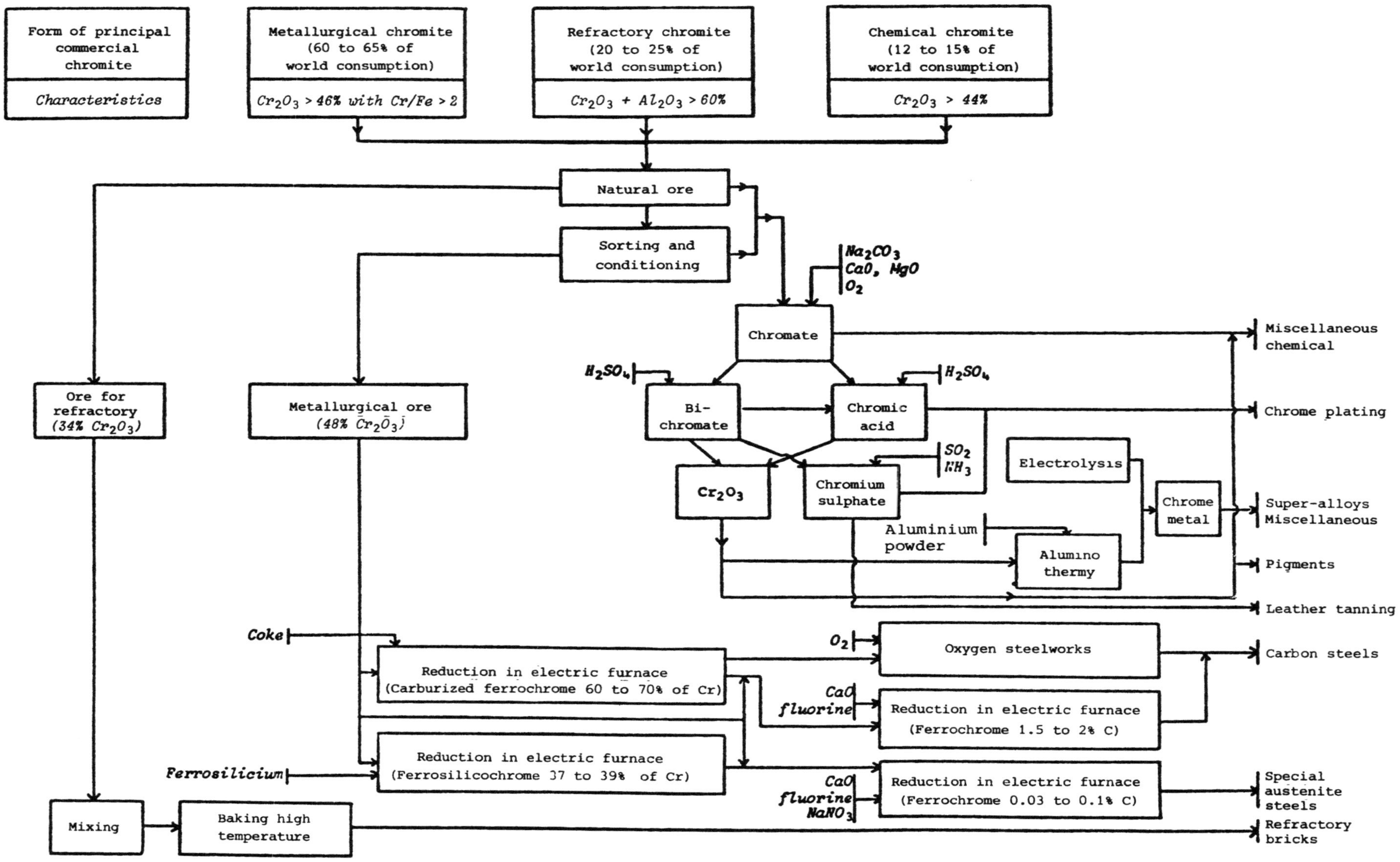

of operation (formation and utilization of basic materials and their derivatives, as well as the effluents produced).

2.2 A SUMMARY OF PRODUCTION AND UTILIZATION

a. Production

In **1971**, the world production of chromites was close to **6.3 million tons**, an increase of 5% on the production in 1970 (6×10^6t).

b. Consumption

Various values, generally in agreement, give the world consumption of chromites (Production + Imports - Exports) without taking into account variations in stocks.

Table 2.2.b-1 summarizes the values of the nine European Common Market countries.

Table 2.2.b-1
Consumption of chromites in thousands of tons for the nine European Common Market countries for 1969, 1970 and 1971 (5,12)

Year \ Country	B/L	RFA	F	UK	I	NL	TOTAL
1969	18	486	330	200	160	4	1198
1970	18	491	421	165	142	6	1243
1971	22	483	340	207	201	10	1253

In comparison, Table 2.2.b-II gives the corresponding figures for Japan, Sweden and United States of America.

Table 2.2.b-II
Consumption of chromites in thousands of tons for 1969, 1970 and 1971 (5,12)

Year \ Country	Japan	Sweden	USA
1969	730	184	1,270
1970	1,149	197	1,260
1971	1,162	188	990

In 1970, the consumption of the United States and of the nine European Common Market countries was very similar, and in 1971 Japan's consumption overtook that of the United States. To complete the report, it is interesting to convert the tonnage of minerals into potentially pure chromium using the following calculations.

. Metallurgical chromites 33%

. Chemical chromites 30%

. Refractory chromites 25%

The average chromium mineral containing 31% chromium.

Taking into account this average mineral content, table 2.2.b-III gives the weight of chromium corresponding to the consumption of minerals mentioned in the above tables.

Table 2.2.b-III
Consumption of chromites and the corresponding chrome contained in thousands of tons for the nine European Common Market countries, Japan, Sweden and the U.S.A.

Mineral / *Cr contained*	The nine European C.M. Countries	Japan	Sweden	USA
1969	1,198 / *370*	730 / *226*	184 / *57*	1,270 / *394*
1970	1,243 / *385*	1,149 / *356*	197 / *61*	1,260 / *391*
1971	1,253 / *388*	1,162 / *360*	188 / *58*	990 / *307*
Average over 3 years	1,230 / *381*	1,016 / *315*	188 / *59*	1,172 / *364*

c. Utilization

For the United States, the breakdown of the tonnage of chromites used in the three main industrial sectors is given below in table 2.2.c-I.

Table 2.2.c-I
Breakdown of tonnages of chromites as thousands of tons, used in the United States of America in the three main industrial sectors (5)

Year \ Sector	Metallurgical	Refractory	Chemical	TOTAL
1969	808	272	190	**1,270**
1970	821 *(48%)*	250 *(36%)*	192 *(45%)*	**1,260**
1971	648	174	162	**990**

(in italics, the Cr_2O_3 *content)*

If the supposition is made that the breakdown for the nine European Common Market countries approximates to the model suggested in a report produced in Great Britain for the year 1970, that:

. 62% of **chromites** are used in **metallurgical** industries (ferro-chromes and other chromium alloys),

. 22% in **refractory** processes and,

. 16% in **chemical** processes

then an equivalent table 2.2.c-II is obtained, for comparison with that given above for the United States.

Table 2.2.c-II
Breakdown of the tonnage of chromites, expressed in thousands of tons, used in the nine European Common Market countries, in the three main industrial sectors

Year \ Sector	Metallurgical	Refractory	Chemical	TOTAL
1969	742 *230*	263 *81*	193 *60*	**1,198** *370*
1970	771 *239*	273 *85*	199 *61*	**1,243** *385*
1971	776 *241*	276 *85*	201 *62*	**1,253** *388*
Average over 3 years	764 *236*	270 *84*	198 *61*	**1,230** *381*

These tables will be supplemented by further statistical information presented later in this report, principally in the more detailed examination of the major uses of the basic metal and metal derivatives, given in the following chapters.

2.3 SOME APPLICATIONS

a. Metallurgical uses

a.1 Chromium steels

The figures which reflect most closely the uses of chromium are those related to the production of ferrochromes, silicochromes and other additions of elements which are principally employed in the formation of stainless steels, refractory steels and tool making processes (small chromium uses are excluded). The primary production figures for these steels for the nine European Common Market countries for 1970 and 1971 are summarized below, units in thousands of tons.

Table 2.3.a-I (11)

Year \ Country	B/L/NL	RFA	F	UK	I	TOTAL
1970	50	503	467	257	237	**1,514**
1971	40	369	406	163	245	**1,223**

For comparison, table 2.3.a-II gives the production of steels, in thousands of tons, for the United States and Sweden.

Table 2.3.a-II (11)

Year \ Country	Sweden	USA
1969	367	1,421
1970	394	1,158
1971	339	1,146

It must be remembered that there is a tendency, especially in the United States, to produce increasingly large quantities of steels containing only very small amounts of chromium. This implies that there is a steady decrease in the amount of steels having high concentrations of added chromium.

a.2 Consumption of ferrochromes, silicochromes and other mixtures of chromium

In 1971, the United States used a little under 300,000 tons of ferrochromes, silicochromes and other mixtures of chromium.

Table 2.3.a-III gives the breakdown of this total into various types of steels and alloys.

Table 2.3.a-III
Consumption in tons of ferrochromes, silicochromes and other mixtures of chromium in the United States for the year 1971 (5)

Types	Low Carbon content ferrochromes	High Carbon content ferrochromes	Ferro silico chromes	Others	TOTAL
Carbon Steels	1,360	3,560	658	-	5,578
Stainless and refractory steels	85,900	63,200	44,600	149	193,849
Alloys other than stainless and refractory steels	15,100	34,050	6,190	2,500	57,840
Tool production steel	435	1,556	172	-	2,163
Cast Iron	655	6,230	93	466	7,444
Special Alloys	7,150	1,070	590	1,705	10,515
Others	4,700	4,620	151	5,790	15,261
TOTAL	115,300	114,286	52,454	10,610	292,650
Chromium contained	*79,100*	*74,000*	*20,400*	*6,560*	*180,060*

a.3 Production of ferrochromes, silicochromes and other mixtures of chromium

Table 2.3.a-IV compares the production in the United States for 1970 and 1971, of ferrochromes, silicochromes and other chromium mixtures (these include those made with chromium metal).

Table 2.3.a-IV
Production in tons of ferrochromes, silicochromes and other chromium mixtures, in the United States for the years 1970 and 1971 (5)

Year \ Type	Low Carbon content ferrochromes	High Carbon content ferrochromes	Ferro silico chromes	Others	TOTAL
1970	116,000	135,500	90,000	14,190	35,5690
Chromium contained	*82,300*	*90,600*	*35,800*	*11,200*	*219,900*
1971	101,000	120,000	83,600	15,800	320,400
Chromium contained	*70,900*	*80,600*	*32,600*	*11,820*	*195,920*

The statistics show that the average content of chromium in ferrochromes, silicochromes and other chromium mixtures used in the metallurgical industries, is 62%.

In addition, the percentage of ferrochromes, silicochromes and other chromium mixtures used in the production of stainless and refractory steels, generally estimated to be 66%, is confirmed by the figures summarized in table 2.3.a-III above (193,849 tons in a total production of 292,650 tons).

It is not possible to obtain such a complete picture for Europe. An estimation based on the two average figures given above is made in order to obtain table 2.3.a-V. This table gives an estimate, for the nine Common Market countries, of the production of chromium and ferrochrome steels.

Table 2.3.a-V
Estimated production, in thousands of tons, of chromium and ferrochrome steels in the nine European Common Market countries for the years 1970 and 1971

Year \ Type	Chromium Steels	Chromium content	Corresponding ferrochromes	Metallurgical ferrochromes
1970	1,514	135	$\frac{135}{0.62} = 220$	$\frac{220}{0.66} = 330$
1971	1,223	112	$\frac{112}{0.62} = 180$	$\frac{180}{0.66} = 270$

These figures are similar to those given in a study in Great Britain in January 1972 (12) for the balance "production + imports - exports" of ferrochromes in Germany, France, Great Britain and Italy for 1970.

b. Chemical production

The most important salt is that of hydrated sodium bichromate $Na_2Cr_2O_7$, $2H_2O$ which is used in the production of:

- chromic acid, principally used in chromium plating processes
- chromium sulphate, principally used in leather tanning
- the anhydride Cr_2O_3, a base for pigments.

"Bichromate equivalent"

Since a large amount of chemical chromites is used in the production of bichromates, which is itself an important base for the formation of many other compounds, it is possible to understand the idea of **"equivalent bichromate"**.

On the basis of the known tonnage of bichromates in the United States and the tonnage of chemical chromites used, as well as the tonnage of the chemical chromites used by the nine European Common Market countries (table 2.2.c-II above), and also the corresponding chromium content tonnage as well as a chemical efficiency of transformation of 80%, table 2.3.b-I gives the tonnage of "equivalent bichromate" used in the chemical sector of the nine European Common Market countries.

Table 2.3.b-I
Tonnage of equivalent bichromate, expressed in thousands of tons, used by the chemical industry of the nine European Common Market countries.

Year \ Products	Chromites used in chemical industry	Total chromium content	Usable chromium content	Equivalent Bichromate
1969	193	60	48	137.5
1970	199	61	48.8	140
1971	201	62	49.6	142

The breakdown for the nine European Common Market countries of the uses of bichromate (and derivatives) is given in table 2.3.b-II.

Table 2.3.b-II
Summary of the various uses of bichromate in the nine European Common Market countries for the year 1971

Uses \ Percentage and tonnage	Percentage	Tonnage in thousands of tons
Chromium plating	25	35.5
Pigments	30	42.6
Leather Tanning	15	21.3
Others	30	42.6
TOTAL	**100**	**142.0**

Taking into account the losses for each industrial sector, these will be examined in paragraph 2.4, and it will then be possible to estimate the corresponding level of pollution.

c. Refractory Industries

The following table shows the tonnage of refractory material produced in the nine European Common Market countries; its principal uses are for ovens in the metallurgical and glass making industries as well as in "linings" for blast furnaces.

Table 2.3.c
Tonnages, expressed in thousands of tons, of chromite used in the nine European Common Market countries for 1969 (12)

Country	RFA	F	UK	I	NL	TOTAL
Tonnage	64	21	68	25	5	183

The figure of 183,000 tons given here approaches that of 263,000 tons, a theoretical value arrived at in table 2.2.c-II above.

It should be noted that there is a diminishing use of chromites in refractory materials destined for the metallurgical industries; this is due to the fact that "Martin" (open hearth) steels are gradually replacing steels produced in oxygen atmospheres.

2.4 AN INITIAL EXAMINATION OF THE QUALITATIVE AND QUANTITIVE WASTES PRODUCED BY VARIOUS INDUSTRIAL PROCESSES DISCUSSED PREVIOUSLY

From chromites, as well as from compounds of the ferrochrome type used in metallurgical processes, salts of chromium used in chemical processes, or oxides of chromium used in the production of refractory materials, the forms of pollution caused in production as well as in the utilization of the products are very varied.

a. **Production losses**

From table 2.2.c-II above, it is possible to obtain already an approximate idea of the size of losses from the various production processes for the nine European Common Market countries.

a.1 In the **production of ferrochromes** complements to thousands are quite variable according to the conditioning and preparation of charges, as to whether dust filters are of low or high efficiency. In the case of low efficiency filters the mineral losses can be as high as 20% for example (some of the dust is in an alkaline form: 40-50% Cr_2O_3; 5-10% CaO).

a.2 Concerning **refractory** materials the only data available is that given by Kienbaum (53) who gives a very small value for losses (1%). This would seem to only apply to installations where recovery of dust and recycling are highly efficient.

a.3 For the **chemical** industry, in order to form the bichromate from the raw mineral, a chemical yield (except in the cases where the installation is very recent or that recovery has been very carefully investigated) of between 80-85% can be expected.

8 to 10% of the chromium is found in the ash "au terril" in an unattacked state.

2% in the waste gases.

The remainder, in the form of Cr^{VI} and Cr^{III}, are for the most part, losses in waste products (this will be discussed further under the section, technical prevention).

If one goes past the bichromate stage, in order to produce chromic acid for example, further losses must be considered.

In order to give a clearer picture, using data based on older types of industrial plant*, the following table is given.

Table 2.4.a-I
Estimated losses of chromium in the three main industrial sectors of the nine European Common Market countries

Sector \ Percentage and tonnage	Percentage	Order of magnitude of losses in elemental chromium (in tons)
Metallurgy	10 to 20	25,000 to 50,000
Refractory	5	4,500
Chemical	15 to 25	9,000 to 15,000

It is evident that the losses shown in the table above have not the same degree of nuisance, in fact, raw minerals are nothing more than losses of earth.

* *According to Kienbaum, the average efficiency of chromites is about 16%. (53)*

b. Losses in cleaning operations in the production of special chromium steels and more specifically, stainless and refractory steels

About 2 million tons of stainless and refractory steels (ingots, moulds and castings), are produced in the world (excluding small chromium) and are cleaned several times in the course of production.

The cleaning is done in acid baths, usually containing nitric acid (HNO_3) and hydrofluoric acid (HF). In reality, it is not so simple, as on one hand, it is not the metals that are attacked but a mixture of complex oxides (Cr_2O_3, CrO, Fe_2O_3, Fe_3O_4, FeO, NiO, etc.) and on the other, the exact composition of the baths is often the result of the working experience of individual operators.

Table 2.4.b-I gives the composition of a cleaning bath for stainless steels, according to Rydberg et al (15).

Table 2.4.b-I

Ions	H+	NO-	F-	Cr	Fe	Ni	MO
Concentration in g/l	1.8	150	30	4	30	7	1

Generally the acid baths are used until the concentration of iron reaches 40 to 80 g/l. However, it should be noted that the waste coming from cleaning operations normally contains only very small quantities of Cr^{VI}.

Table 2.4.b-II gives the amount of chromium in tons, contained in waste effluents produced in cleaning operations of stainless and refractory steels, in the nine European Common Market countries and in Sweden.

Table 2.4.b-II
Production of stainless and refractory steels (ingots, moulds and castings), expressed in tons, in the nine European Common Market countries and in Sweden for the year 1972, and also an estimate of the amount of chrome in waste products from cleaning operations

Country / Production and quantity in tons	B/L/NL	RFA	F	UK	I	Sweden (15)
Production of stainless steels	-	560,000	489,000	225,700	265,800	387,000
Quantity of chromium in waste products	-	320	280	130	150	220

c. Losses on utilization

Turning more specifically to the four major industrial uses of chromium employed by the chemical industry, Table 2.4.c-I shows the associated losses in wastes.

Table 2.4.c-I
Attempts to evaluate the overall losses in wastes of chromium for the chemical industries of the nine European Common Market countries

Input in tons	Principal Uses	Output losses in waste in tons *(1)*	
(2) 35,500 *(3)* *12,400*	Electro deposition (chroming, plating)	26,600 *9,300*	≃ 75%
21,300 *7,460*	Leather Tanning	5,330 *1,860*	≃ 25%
42,600 *14,350*	Pigments	2,130 *745*	≃ 5%
42,600 *14,350*	Various (Chemical oxidation, colouring, pharmaceutical catylizers)	4,260 *1,490*	≃ 10%

d. Other sources of losses to the environment

Some other sources of wastes containing chromium are set out below:

- waste products from the titanium industry (beach sands - ilmenites- can contain chromite),

- waste products from the phosphate industry,

(1) *Without treatment*
(2) *Calculated as equivalent bichromate*
(3) *Calculated as elemental chromium*

- waste products from certain pyrites used in the production of sulphuric acid,

- waste products from the burning of certain fossil fuels (coal, petrol),

- waste products produced by the oxidation of organic material by chromic compounds.

2.5 AN EXAMINATION OF VARIOUS KNOWN PREVENTATIVE TECHNIQUES FOR WASTE PRODUCTS CONTAINING CHROMIUM

In paragraph 1.3.b.1. of chapter I, a preliminary discussion of the principal types of pollution coming from industrial uses of chromium was undertaken. In point 2.4 of the present chapter, preliminary information has been given on the wastes being produced by individual processes or operations within the chromium industry.

It is now necessary to discuss in more detail the various preventative techniques applicable. Figure 2.5.I below, summarizes the main chemical processes used in transporting chromites, valence III to valence state VI, and into metallic form.

For each step in the transformation from one chemical form to another, there are losses of chromium, either in hexavalent or trivalent states. Treatment techniques depend upon considerations, shown below, which are not only applicable to production operations, but also for industrial users of chromium.

a. Hexavalent chromium (Cr^{VI})

a.1 Introduction

In industrial effluents, hexavalent chromium is often found in the ionic form CrO^4 (chromate) or Cr^2O^7 (bichromate). Although the most important sources of pollution are generally from wastes containing chromic acid, from processes of deposition and electro-plating being mainly practised in the automobile industry, hexavalent chromium can also arise form:

- industrial water cooling (refrigeration) circuits to which are added certain chromium compounds that act as corrosion inhibitors,

- the production of pigments

- leather tanning

- aluminium anoding (galvanising)

- polishing and surface treatments

FIGURE 2.5.I - DIAGRAM OF THE MAIN CHEMICAL PROCESSES IN THE TRANSFORMATION OF CHROMITE, VALENCE III TO VALENCE VI, AND INTO METALLIC FORM

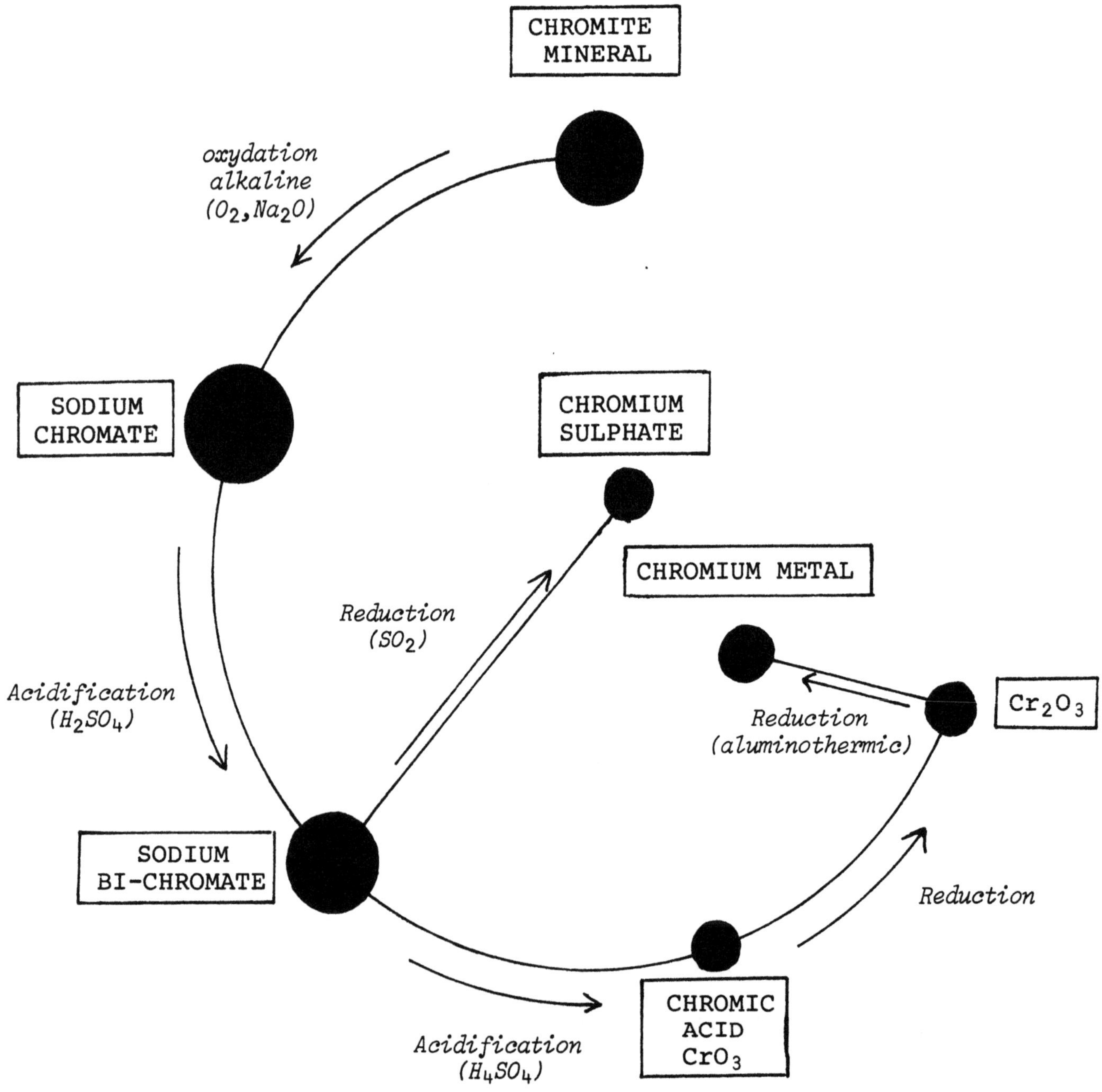

Table 2.5.a-I gives some examples, based on an American study (49), of industrial sources of effluents containing hexavalent chromium as well as concentrations.

Table 2.5.a-I
Some industrial sources of effluents containing Cr^{VI} in the United States (49)

Source of effluents	Concentration of Cr^{VI} in effluents mg/l	
	Average	Extreme limits
Leather tanning	40	
Wood treatment		0.2 - 1.5
Water after passing through industrial cooling systems	30	8 - 60
Polishing baths		200 - 75,000
Rinsing water from polishing operations		1 - 6
Galvanizing baths		15,000 - 75,000
Rinsing water from galvanizing operations	50	30 - 100
Plating		60 - 270,000
Electro plating	40	15 - 70

Various treatment techniques eliminating hexavalent chromium in industrial effluents are in use:

- reduction and precipitation

- ion exchanges

as well as other methods such as:

- concentration by evaporation

- adsorption on active charcoal (carbon)

which may prove useful and are in the course of being studied.

a.2 Reduction and precipitation

This method consists essentially of lowering the pH value to about 3, by the addition of sulphuric acid, and then adding a reducing agent such as SO_2, $NaHSO_3$, $FeSO_4$, $Na_2S_2O_5$ ...

Examples of such reducing reactions are set out below:

$$3SO_2 + 2H_2CrO_4 \rightarrow Cr_2(SO_4)_3 + 2H_2O$$

$$6NaHSO_3 + 3H_2SO_4 + 4H_2CrO_4 \rightarrow 2Cr_2(SO_4)_3 + 3Na_2SO_4 + 10H_2O$$

$$6FeSO_4 . 7H_2O + 6H_2SO_4 + 2H_2CrO_4 \rightarrow Cr_2(SO_4)_3 + 3Fe_2(SO_4)_3 + 15H_2O$$

The reaction of sodium meta bisulphate is identical to that of bisulphate:

$$Na_2S_2O_5 + H_2O \rightarrow 2NaHSO_3$$

It should be noted here that an excess of SO_2 will cause an oxygen demand, if proper aeration has not taken place in the course of the reaction.

The trivalent chromium that is thus formed is then eliminated by precipitation by the addition of CaO and finally by polyelectrolytes used as floculating agents.

These precipitation reactions are shown below:

$$Cr_2(SO_4)_3 + 3Ca(OH)_3 \rightarrow 2Cr(OH)_3 + 3CaSO_4$$

or

$$Cr_2(SO_4)_3 + 3Fe_2(SO_4)_3 + 12Ca(OH)_2 \rightarrow 2Cr(OH)_3 + 6Fe(OH)_3 + 12CaSO_4$$

Thus by reduction, at a pH of between 2.5 and 4, it is possible to obtain a residual concentration of hexavalent chromium of about 0.01 mg/l.

a.3 Waste treatment by ion exchange

- In general anion type ion exchange resins are used in the treatment of effluents containing hexavalent chromium and cation ion exchange resins used for effluents containing trivalent chromium.

  These techniques become economic once the chromate level in the effluent is greater than 200 mg/l.

In general, treatment techniques are dependent upon the concentration in the effluents, thus:

- ion exchange below 200 mg/l,
- precipitation between 200 mg/l and 500 mg/l,
- evaporation above 500 mg/l.

As an example of the elimination of heavy metals, in the absence of cyanides, figure 2.5.a-II gives a schematic diagram of a treatment installation for a hard chromium process workshop (30).

. It is also possible to recover chromic acid from rinsing operations of electroplating by using cationic resins in the hydrogen form; the cations present such as Na, Fe, Sn, Cu and Cr are eliminated at the same time, the ion exchange column step being followed by a single or multiple evaporation step.

It is thus possible to recover chromic acid (about 50 g/l) at a cost of 0.20 u.c. per kg, the free market price being in the order of 0.60 u.c. per kg*.

a.4 Concentration by evaporation

This treatment is applied to rinsing waters of plating processes. It consists of selective evaporation of the water in the plating bath in order to concentrate the chromic acid so that it can be recovered.

a.5 Adsorption on active carbon (charcoal)

This procedure, still undergoing tests, seems to promise a means of treating effluents containing hexavalent chromium in only fractions of mg/l up to several hundreds of mg/l. The efficiency of this treatment process would be greater than 50% and could reach 99% in some cases.

a.6 Concentration by freezing

This procedure, still only in the research stage, although having been tested with some success on pilot plants in America, is used on effluents from plating processes.

The materials thus concentrated by freezing may be re-used directly, or can be further concentrated by precipitation.

Table 2.5.a-III summarizes the results of the various treatment techniques in actual use, or in the research stage.

**Private communication*

FIGURE 2.5.a-II - SCHEMATIC DIAGRAM OF A TREATMENT INSTALLATION FOR A HARD CHROMIUM PROCESS WORKSHOP

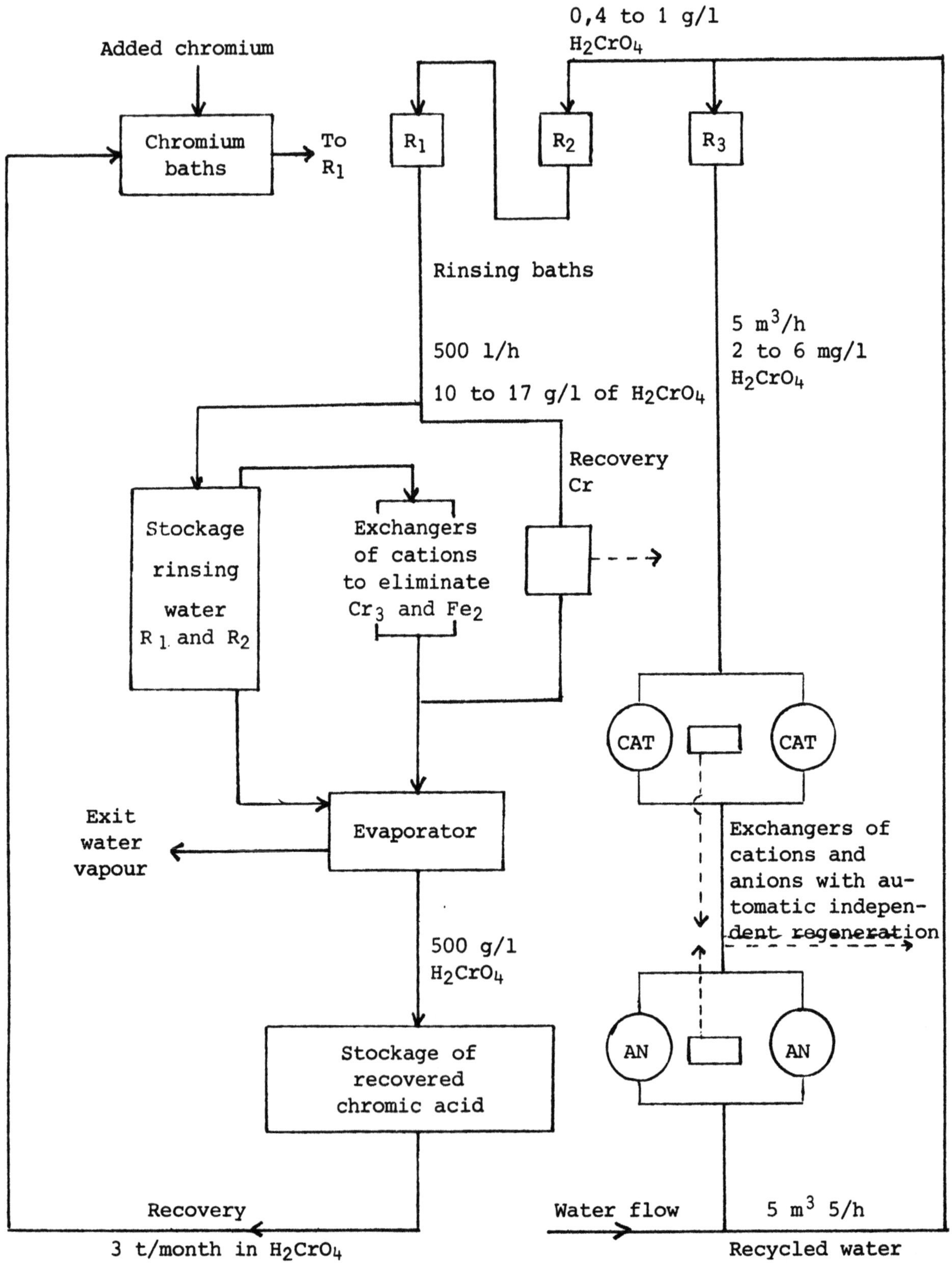

Table 2.5.a-III
Results from various treatment processes for effluents containing hexavalent chromium (49)

Treatment processes	Concentration of hexavalent chromium in effluents	
	initial	final
Reduction		
SO_2	1300	
Bisulphite	140	0.001 to 1
Metabisulphite	70	
Ion exchange	8 - 10	0 - 0.9
Active carbon	0.09 - 820	0.006 - 195
Concentration by evaporation	100	0.2

b. Trivalent chromium (Cr^{III})

b.1 Introduction

Trivalent chromium found in industrial effluents can come from industries using directly Cr^{III} such as:

. glass manufacturing,

. production of ceramics

. photographic industries,

. textile colouring manufacturers,

or from waste treatment plants in which primary effluents contain hexavalent chromium which is transferred to the trivalent state during the course of treatment.

In addition, in certain residues of electro-plating processes, chromium may be found in the trivalent form before any reduction has taken place; concentration may range from a few g/l up to several hundred g/l of Cr^{III}.

Two main treatment processes exist for effluents containing trivalent chromium:

- precipitation,

- ion exchange.

b.2 Precipitation

The method consists essentially in forming insoluble chromium hydroxide, $Cr(OH)_3$, by the addition of sodium hydroxide to the treated effluents.

The optimum conditions for the precipitation is a pH between 8.5 and 9.5, which corresponds to the minimum solubility of chromium hydroxide. In addition, certain floculating agents may be employed in order to solidify the precipitate and to cause sedimentation of $Cr(OH)_3$, thus increasing the efficiency of the treatment.

The residual gelatinous precipitates are often very difficult to solidify and various methods may be employed to collect them. These may consist of simple filtration, vacuum filtration and even centrifugation, but more often, the common procedure is to pour the precipitates into a large open air receptacle and allow setting.

b.3 Ion exchange

Because of the electric properties of Cr^{III} (positively charged), cation exchange resins must be employed. (see also point 2.5.a.3 above).

The technique has the advantage that it eliminates, at the same time as trivalent chromium, other heavy metals such as iron and aluminium, thus purifying chromic acid so that it may be recycled. However, this acid can attack the resin itself, if its concentration is greater than a certain level ($\simeq$ 250 g/l), thus it is often necessary to dilute the effluents before purification, the chromic acid recovered having to be reconcentrated before re-use.

Although this method is little used at present in industry, ion exchange techniques applied to effluents containing Cr^{III} have a well established efficiency and are able to eliminate nearly all trivalent chromium in effluents treated in such a manner.

c. Industrial examples

Table 2.5.c-I below gives a breakdown of industrial sources and approximate quantities of waste products in France (hexavalent chromium plus trivalent chromium expressed in tons of Cr metal).

Table 2.5.c-I

Industrial source	Waste Estimated in tons of Cr metal
Surface treatment	1,700
Pigments	10
Dyeing	5
Oxidation	500
Leather tanning	500
TOTAL $Cr^{VI} + Cr^{III}$	≃ 2,800

The schematic diagram given in figure 2.5.c-II below, shows an example of anti-pollution treatment with recycling found in the production of exchromite chemical intermediaries.

Three treatment procedures may be used, their major characteristics are outlined below:

- treatment by reduction, floculation, decantation and filtration of dilute chromium containing wastes,

- treatment by ion exchange resins or organic solvents of effluents containing chromium at medium concentrations,

- treatment by solvents of effluents of high chromium content.

Figure 2.5.c-III gives the balance of Cr^{VI} and Cr^{III} coming from effluent purification operations according to the various combinations of processes shown below.

c.1 Treatment by reduction, floculation, decantation and filtration of dilute chromium containing wastes

(Treatment of polluted waters containing hexavalent and trivalent chromium)

1) Nature of the effluents containing Cr^{VI} < 0,5 g/l

Effluents containing more than 0,5 g/l Cr^{VI} may be treated by this method but it is, however, economically and ecologically more efficient to treat them by one of the two other methods mentioned hereafter.

FIGURE 2.5.c-II - EXAMPLE OF ANTI-POLLUTION TREATMENT WITH RECYCLING FOUND IN THE PRODUCTION OF EX-CHROMITE CHEMICAL INTERMEDIARIES

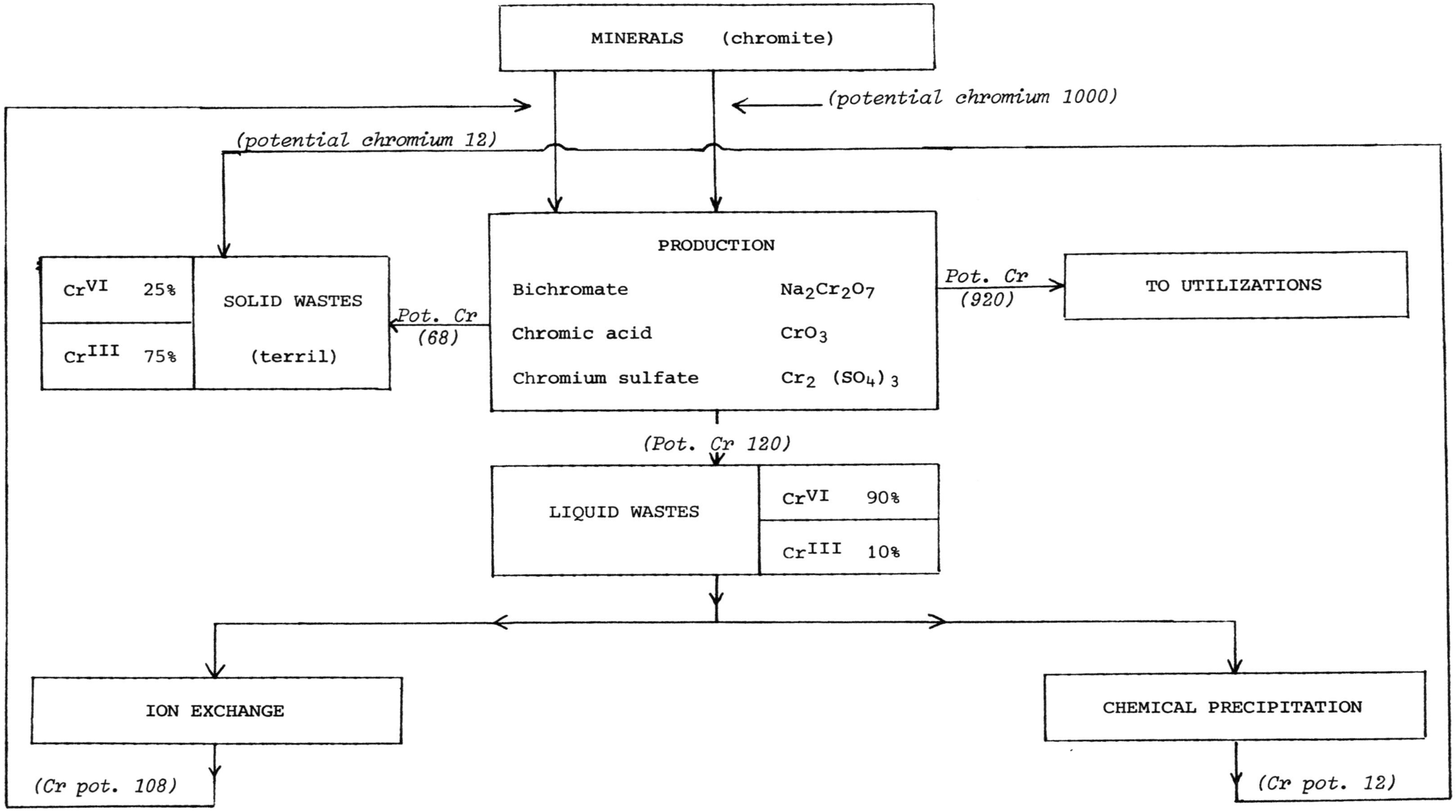

FIGURE 2.5.C-III – PURIFICATION OF CHROMIUM EFFLUENTS (balance of [illegible] and of [illegible] expressed in g [illegible])

A - Purification by solvants

B - Purification by ion exchange resins

C - Purification by physico-chemical breakdown (reduction, floculation, decantation, filtration)

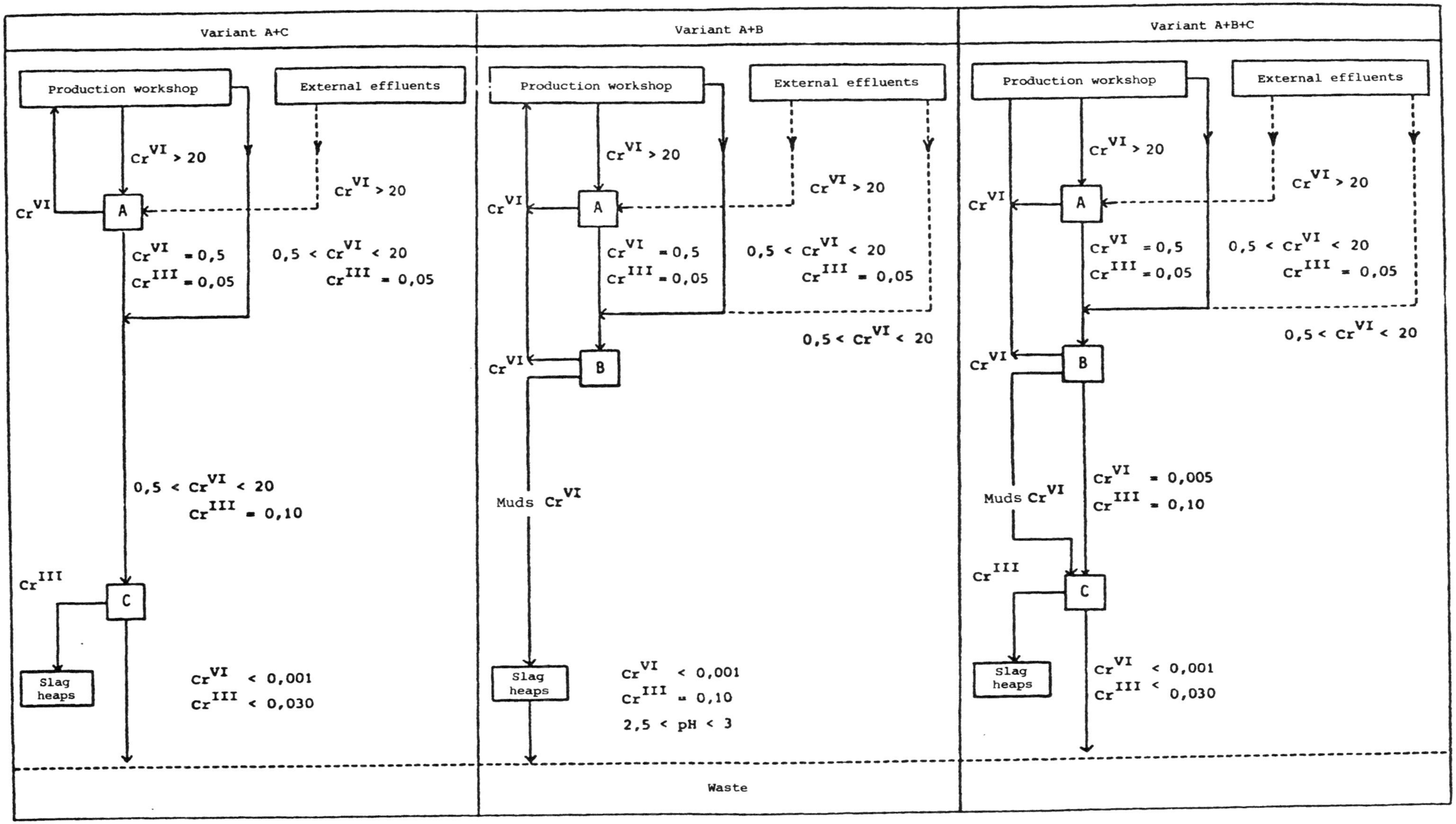

* *Private Communication*

NOTE – Phase C only could be used, which would mean establishing a pH of 2.5, a reduction of CrVI to Cr III by adding SO_2 for example, then neutralizing in one or more stages by adding lime up to pH 8. The addition of small quantities of approximately the fraction of ppm of anionic polymerics, aids floculation. The muds are decanted which can be concentrated into dry materials by centrifugalizing.

2) Concentration of toxic materials in the waste after treatment

$Cr^{VI} < 0,0001$ g/l

3) Summary of method

A summary of the chemical method of reduction of hexavalent chromium to trivalent chromium and its elimination in this form by floculation-decantation and filtration of the precipitate.

4) Economic and ecological advantages of the method

Assures the total elimination of Cr^{VI} but produces a solid Cr^{III} effluent.

5) Effect on this method when treatment plants are close to the sea

In the case where treatment plants have effluent residues passing to purification stations (sewage works) and which are particularly biological in nature, care must be taken as chromium acts as an inhibitor.

Although certain wastes may reach the sea directly, the pH of seawater is such that hexavalent chromium should not exist, but phenomena such as precipitation and eventual complexation and the effect of the marine bios do not seem to have been studied in detail at present.

c.2 Treatment by ion exchange resins or organic solvents of effluents containing chromium at medium concentrations

(Treatment of polluted waters containing hexavalent chromium and changing this into chromite sulphate for leather tanning)

1) Nature of wastes containing $Cr^{VI} \simeq 0,5$ g/l and having an absence of chlorine, large volumes of suspended material (muds, solid waste) and organic matter

2) Concentration of toxic materials in waste after treatment

$Cr^{VI} < 0,001$ to $0,0001$ g/l

3) Summary of method

Treatment of residue solutions containing compounds of hexavalent chromium with a view to rendering them harmless and producing at the same time chromium sulphate by extracting hexavalent chromium, either by an ion exchange resin or by an organic solvent insoluble in water and then re-extracting the chromium using a reducing agent (for preference sulphur dioxide in the presence of sulphuric acid).

This process is covered by a patent*n° 72-13317 made on the 14th April 1972 in France, Great Britain, Belgium, Spain, Holland, Italy, Luxembourg, Norway, Federal Republic of Germany, Switzerland, Sweden, United States of America and Japan.

4) Economic and ecological advantages of the method

In contrast to other known methods of detoxification of chromium containing effluents, the present technique produces no solid waste. Moreover, it allows a recycling of chromium which is favourable from the point of view of the environment.

c.3 Treatment by solvents of effluents of high chromium content

(Treatment of polluted waters containing hexavalent chromium and changing this into sodium bichromate)

1) Nature of wastes containing $Cr^{VI} > 0,5$ g/l

2) Concentration of toxic materials in waste after treatment

$Cr^{VI} < 0,5$ g/l

This waste can then be treated by the ion exchange resin method discussed above in c.2.

3) Summary of method

Treatment of residue solutions containing compounds of hexavalent chromium with a view to rendering them harmless and producing at the same time sodium chromate or bichromate by extracting the hexavalent chromium with insoluble organic solvent and re-extracting the chromium from this solvent by an aqueous solution.

This process is covered by a patent* n° 72-13318 made on the 14th April 1972 in France, Great Britain, Belgium, Spain, Holland, Italy, Luxembourg, Norway, Federal Republic of Germany, Switzerland, Sweden, United States of America and Japan.

4) Economic and ecological advantages of the method

In contrast to other known methods of treatment of chromium containing effluents, the present technique produces no solid waste. Moreover, it allows a recycling of the chromium which is favourable from the point of view of the environment.

d. General economic information on the treatment of chromium containing effluents

It should be noted that the chemical products used in treatment operations as mentioned above, are reasonably inexpensive to obtain.

**Société des Produits Chimiques Ugine Kuhlmann*

Table 2.5.d-I gives an idea of the prices in the United States in 1970 for those chemicals used in the treatment of wastes containing hexavalent chromium (16).

Table 2.5.d-I
An indication of the price of chemicals needed for the treatment of effluents containing hexavalent chromium, expressed in u.c. per kg of Cr^{VI} reduced (16)

Process	Ferrous Sulphate	Metabisulphate	SO_2
$FeSO_4 \cdot 7H_2O$ / *24.5 u.c/t*	16 kg / *0.39 u.c*	-	-
H_2SO_4 (66° baumé) / *32.2 u.c/t*	6 kg / *0.2 u.c*	1.5 kg / *0.05 u.c*	-
$Ca(OH)_2$ / *19.6 u.c/t*	10.5 kg / *0.2 u.c*	2.6 kg / *0.05 u.c*	2.6 kg / *0.05 u.c*
$Na_2S_2O_5$ / *12 u.c 100 kg* $NaHSO_3$	-	2.9 kg / *0.35 u.c*	-
SO_2 / *0.11 u.c/kg*			1.9 kg / *0.21 u.c*
TOTAL COST	0.8 u.c	0.45 u.c	0,26 u.c

It is thus obvious that if these prices are low then they represent only a fraction of the total treatment costs. This cost is difficult to determine generally, as it depends greatly upon the dilution of the effluents, the levels that will be required by regulations now in the process of being elaborated, and the fact that the value of the recovered elements is tending to increase.

Lastly, table 2.5.d-II gives an idea of the investment costs for a small chromium treatment plant.

Table 2.5.d-II*

Treatment capacity	Estimated cost of investment	Running costs	Installation power
2 m^3/h	25,600 u.c°	0.08 to 0.16 u.c/m^3	5 kW
20 m^3/h	28,800 u.c° (automatic units)	-	-
6 m^3/h	12,300 u.c°° (manual unit) 19,500 u.c°° (automatic unit)	-	-
2,5 m^3/h	4,650 u.c°° (manual unit) 12,000 u.c°° (automatic unit)	-	-

**Private communication*

° *1974 costs*
°° *1971 costs*

CHAPTER 3

ORIGINS, FORMS AND CHARACTERISTICS OF ZINC AND CADMIUM POLLUTION

CHAPTER 3 - ORIGINS, FORMS AND CHARACTERISTICS OF ZINC AND CADMIUM POLLUTION

3.1 SOURCES AND PRODUCTION

a. Types of minerals

a.1 Zinc

It is worthwhile recalling that the main form of zinc found in nature is as **sulphide complexes** (blends), or **silicate complexes** (calamine) and also as **carbonates.**

The blend (ZnS) often contains iron (up to 26%), some manganese, **cadmium** (up to 5% cadmium as in the case of libramite), silver and more rarely tin and **mercury**, this last element being often found in treated effluents. The zinc blend is regularly found associated with galenite (PbS) which is the principal mineral of lead and which also contains other elements such as zinc and cadmium.

Calamine is a silicate complex **(SiO_2, 2ZnO, H_2O)** and is largely associated with other minerals of zinc.

Smithsonite or **zinc carbonate (CO_3Zn)** may also contain certain other elements such as Fe, Mn, Ca, Mg and more rarely Cd and In.

a.2 Cadmium

Cadmium is normally very closely associated in nature with zinc.

Found only in small concentrations, cadmium represents approximately 10^{-5}% of the earth's crust. Cadmium is not found in a free state in nature and its only known mineral is **Greenockite (CdS).**

It should be recalled that in the production of 1 ton of zinc, 2,5 to 3 kg of cadmium is produced.

b. Production of the metal and metal derivatives

Because of the nature of the minerals used it is difficult to separate the metallurgy of zinc and cadmium from that of lead.

Figure 3.1.b-I (see below), shows schematically the industrial sources and treatment of lead and zinc and also the inter-relationships of the metallurgy of the three elements.

FIGURE 3.1.b-I - INDUSTRIAL SOURCES AND TREATMENT OF LEAD AND ZINC *(Private communication)*

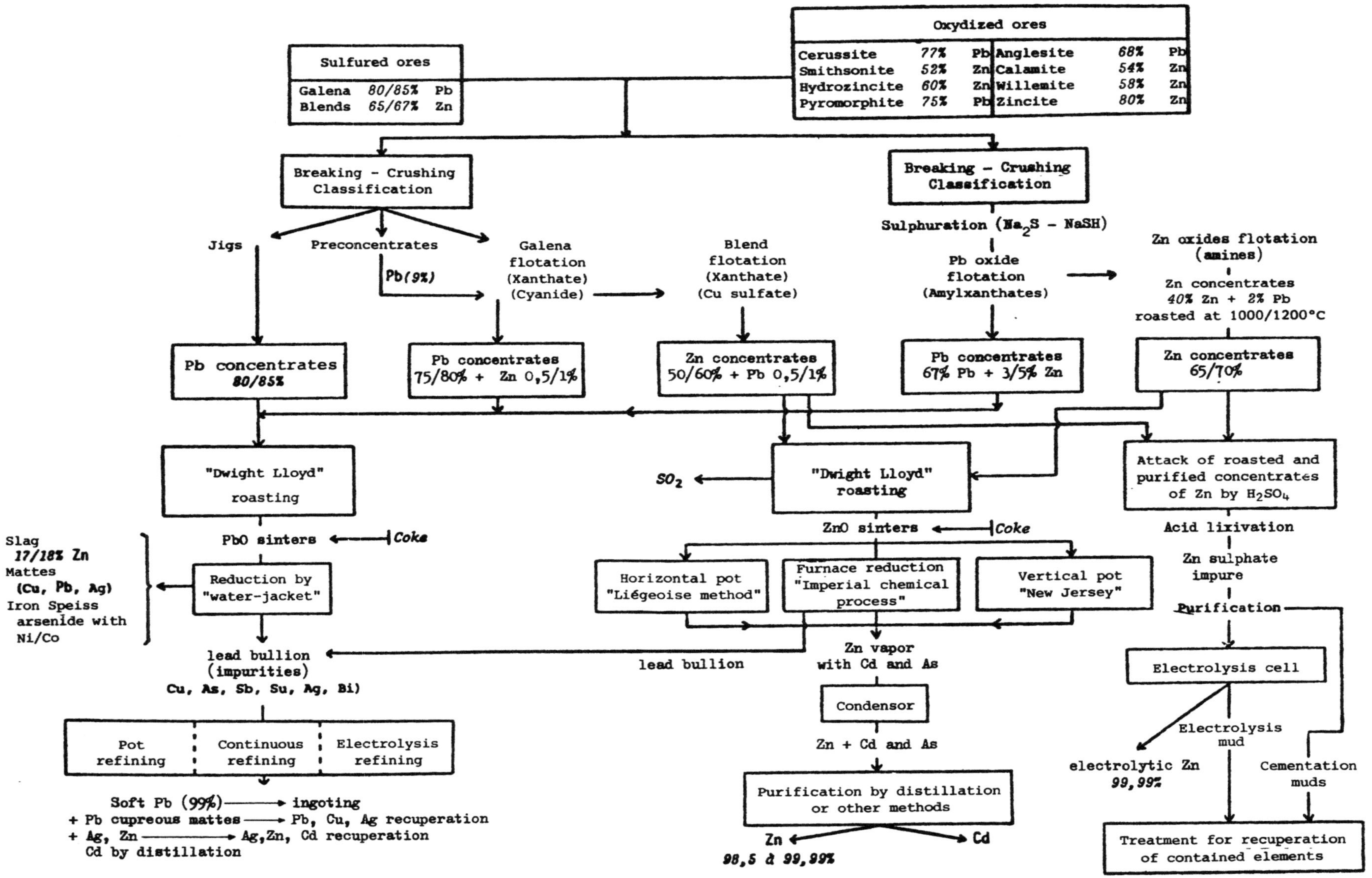

In Annex 2 at the end of this report, several simplified synoptic diagrams are given showing the forms of pollution associated with each stage of the production and use of zinc and cadmium.

Figures 3.1.b-II and 3.1.b-III below, show more precisely the various stages involved in the metallurgy of zinc in both wet and thermal processes, as well as the major production points of associated pollutants.

3.2 SUMMARY OF PRODUCTION AND UTILIZATION

a. Zinc industry

a.1 Production

1) Mineral production

The production of minerals is generally given in terms of zinc content but can also be given as recoverable zinc. Several values are available of which the main ones are summarized and compared below.

Thus in 1966 to 1972, the world production of zinc minerals is given in table 3.2.a-I, taken from a German reference (4).

Table 3.2.a-I
World production in thousands of tons

Year	Zinc content	Recoverable zinc
1966	4,442	4,120
1967	4,733	4,385
1968	4,911	4,545
1969	5,228	4,835
1970	5,352	4,955
1971	5,370	4,985
1972	5,486	5,080

From these figures it is immediately possible to see that between 1970 and 1972, theoretical losses were in the order of 8% of zinc content and real losses were of the same magnitude.

FIGURE 3.1.b-II - ZINC METALLURGY (thermal process) (19, 22, 44)
Following unitarian operations including the recovery of cadmium

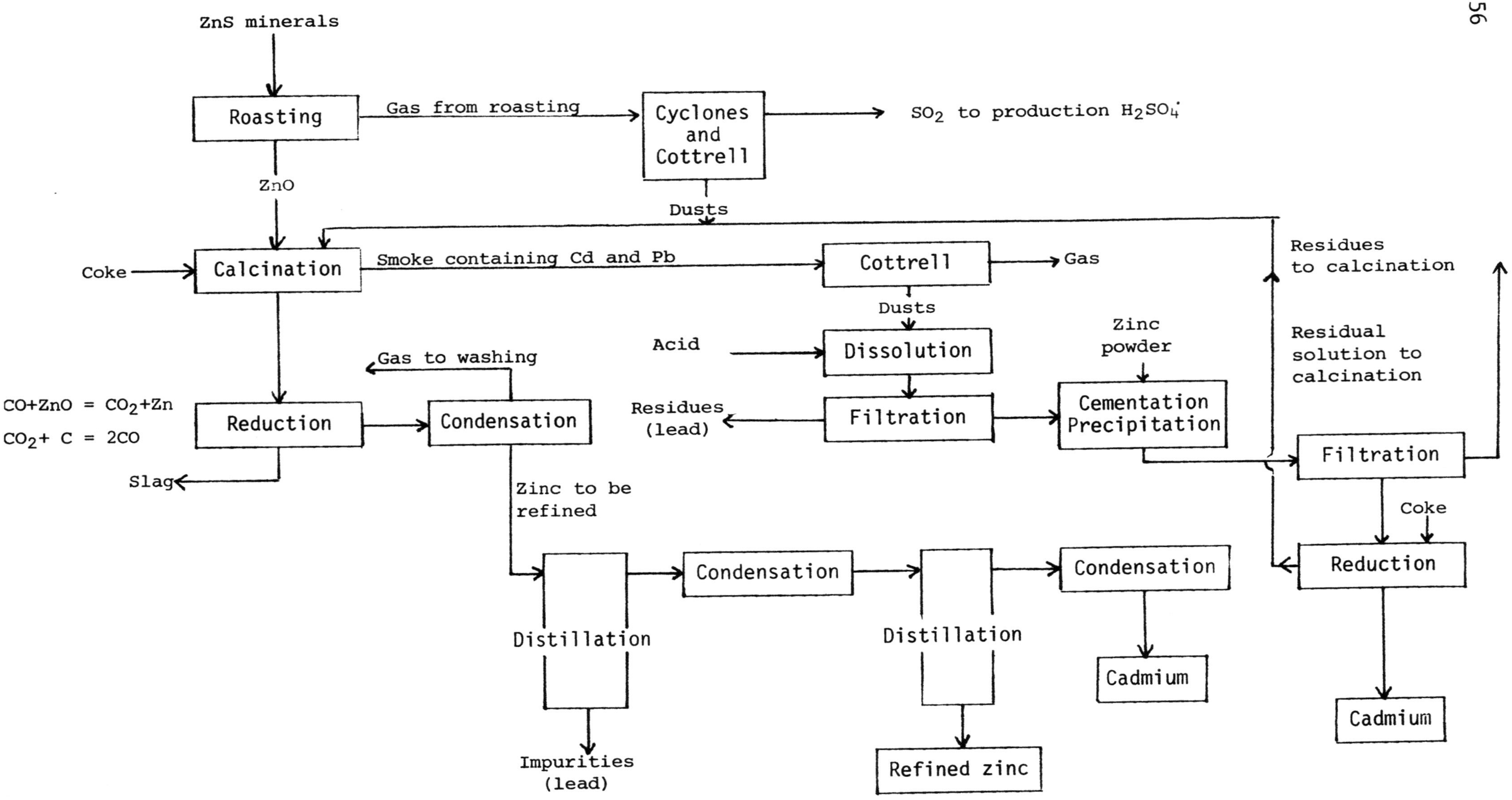

FIGURE 3.1.b-III - ZINC HYDROMETALLURGY (wet process) (19, 22, 44)

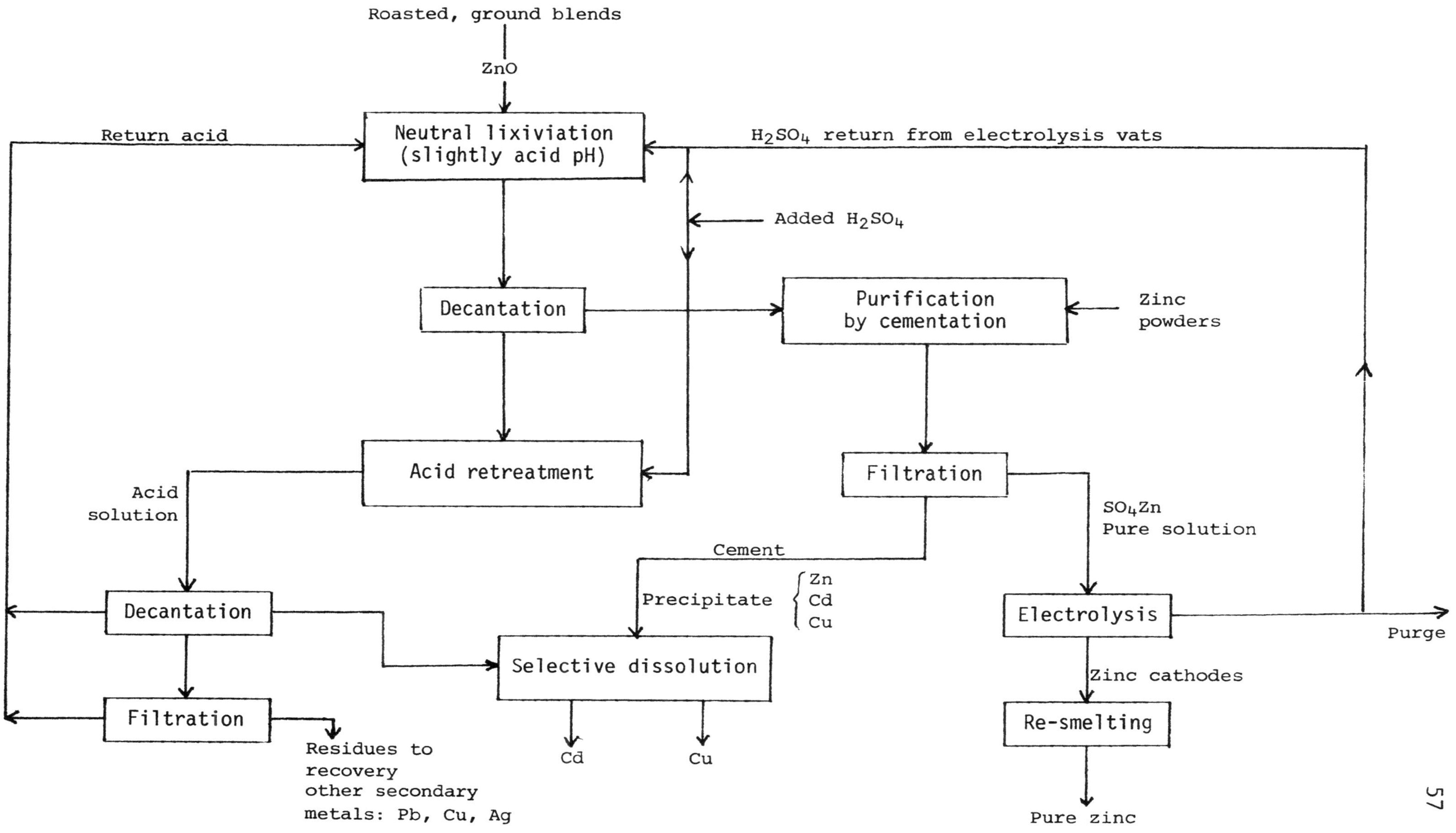

Table 3.2.a-II gives the world production of zinc minerals between 1967 and 1971, as shown by American statistics.

Table 3.2.a-II
World production in thousands of tons (3)

Year	Mine production (zinc content)	Foundry production (metal zinc)
1967	4,830	4,120
1968	4,970	4,630
1969	5,340	4,970
1970	5,450	4,880
1971	5,510	4,610

In general the figures given under the column heading **"Foundry production"** show the amount of zinc obtained from domestic and imported minerals plus zinc recovered from metallic waste (scrap).

It is apparent, from the statistics available, that the annual recoverable zinc from minerals has been more or less constant from 1970, and is about 5,000,000 tons.

2) Comparison of the production of the states members of the European Economic Community with that of the United States of America

The minerals available for the states members of the European Economic Community, including domestic production, increased by imports and decreased by exports, was in the region of 1,000,000 tons zinc content for 1970.

For this year, the main figures are given below (in tons) (4)

Country	Production	Availability
Belgium/Luxembourg	0	205,000
France	19,000	254,000
United Kingdom	0	154,000
Italy	111,000	104,000
Federal Republic of Germany	138,000	291,000
TOTAL	**260,000**	**1,008,000**

The final figure (1,008,000 t) gives an idea of the importance of zinc (and of cadmium) in Europe.

For comparison the corresponding value for the United States of America in 1970 was 1,010,000 tons.

It is also possible to compare the production of zinc metal, this comparison being, however, more difficult, as the figures given in the statistics may or may not include the tonnages of zinc coming from waste recovery processes (distillation of primary scraps, treatment of secondary recoveries).

Table 3.2.a-III gives the yearly figures of production of zinc metal between 1967 and 1971 for the major producing countries in the European Economic Community, values taken from an English reference source (8).

Table 3.2.a-III
Production of zinc metal in thousands of tons for the nine member states of the European Economic Community (8)

Year \ Country	B/L	RFA	F	I	NL	UK	TOTAL
1967	227	180	187	88	39	104	825
1968	250	202	207	113	49	142	963
1969	260	280	253	131	47	151	1,122
1970	241	300	224	142	47	146	1,100
1971	213	253	218	140	41	116	981

For the same period, the production of the United States of America is shown in table 3.2.a-IV, values taken from an American reference source (10).

Table 3.2.a-IV
Production of zinc metal in thousands of tons (USA)

Year	Zinc metal produced from mineral sources	Zinc metal produced from scraps	Total
1967	844	66	910
1968	918	72	990
1969	937	63	1,000
1970	799	67	866
1971	690	67	757
1972	629	61	690

Until 1969, the production figures for zinc metal, ex-ore, in Europe and the United States of America were fairly comparable, but from 1970-1972 production in America progressively reduced by about 30%.

a.2 Consumption

From the point of view of the protection of the environment, the figures relating to the world consumption of zinc, of all types, are very important.

Table 3.2.a-V, taken from a German reference (4), shows the consumption of slabs of zinc between 1967 and 1972, for the member states of the European Economic Community.

Table 3.2.a-V
Consumption of slabs of zinc in thousands of tons for the member states of the European Economic Community.

Year \ Country	B/L	RFA	DK	IRL	F	UK	I	NL	TOTAL
1967	119	303	8	4	202	259	141	30	**1,066**
1968	122	361	10	4	202	281	155	38	**1,173**
1969	152	398	12	5	239	289	167	34	**1,296**
1970	137	396	12	3	220	278	178	37	**1,261**
1971	136	387	9	3	225	274	170	37	**1,241**
1972	145	413	13	3	264	277	203	37	**1,355**

The figures show an increase in consumption in the member states of the European Economic Community of about 30% in five years.

For the United States, the consumption is comparable with that of Europe although the increase is smaller during the same period.

Table 3.2.a-VI
Consumption of slabs of zinc in thousands of tons (10)

Year	1967	1968	1969	1970	1971	1972
Consumption	1,130	1,221	1,252	1,104	1,137	1,295

Lastly, table 3.2.a-VII shows an estimation of the consumption of zinc in 1972 for the rayon industry in Western Europe, (European Economic Community and European Free Trade Association).

Table 3.2.a-VII

Type of rayon	Tonnage production (in tons)	Consumption of zinc unit (kg of Zn/t)	Consumption of zinc total (in tons)
Rayon yarn	140,000	30	4,200
High resistance rayon	135,000	12	200
Rayon skein	600,000	10	6,000
TOTAL	875,000		10,400

From an examination of the statistics it becomes obvious that changes in short term values for consumption are difficult to use for making deductions on long term tendencies, because of factors such as: storage and depletion of stocks, in which changes occur during various periods of crisis. To these factors can be added, as well, the distortions produced by operations associated with the recovery of zinc from scraps.

During the period 1960 to 1970, the consumption of zinc has increased at a rate of approximately 6% per year, from 3,000,000 tons to 5,000,000 tons.

However, in the period 1970 to 1975, it is reasonable to expect an increase in capacity of about 500,000 tons, the greater part being due to the replacement of older processes by more modern methods (Imperial Smelting and electrolytic processes).

b. Cadmium industry

b.1 Production

Table 3.2.b-I gives the production figures for cadmium in six member states of the European Economic Community, statistics taken from a German reference (4).

Table 3.2.b-I
Production of cadmium in the United States in tons

Year \ Country	B/L	RFA	F	UK	I	NL	TOTAL
1967	815	398	499	209	218	107	2,246
1968	819	342	552	205	250	112	2,280
1969	949	792	523	245	422	111	3,042
1970	1,095	1,035	528	318	426	111	3,513
1971	947	982	579	262	350	123	3,243
1972	1,150	913	572	242	416	122	3,415

For comparison, the following tables give, over the same period, the production of cadmium in the United States of America and world production. The values for the United States (statistics taken from the same source as above), include cadmium obtained from waste recovery processes.

Table 3.2.b-II
Production of cadmium, in tons, in the United States of America

Year	1967	1968	1969	1970	1971	1972
Production	3,046	4,831	5,736	4,293	3,597	3,760

Table 3.2.b-III
World production of cadmium in tons

Year	1967	1968	1969	1970	1971	1972
Production	13,566	14,823	17,333	16,400	15,298	16,524

The world production statistics given by an English reference (6) are slightly higher than those given above, as it is very difficult to avoid counting tonnages twice.

b.2 Consumption

Tables 3.2.b-IV and 3.2.b-V given below, give the consumption figures in seven member states of the European Economic Community and also in the United States of America. Statistics taken from a German reference source (4).

Table 3.2.b-IV
Consumption of cadmium in the nine European Common Market countries (in tons)

Year \ Country	B/L	RFA	DK	F	UK	I	NL	TOTAL t
1967	552	1,420	22	925	1,368	260	32	4,579
1968	608	1,702	24	874	1,439	300	49	4,996
1969	683	2,296	23	1,217	1,473	350	43	6,087
1970	781	1,801	24	988	1,313	370	72	5,349
1971	703	1,788	17	969	1,176	320	108	5,081
1972	1,086	1,964	20	1,000	1,354	350	94	5,868

Table 3.2.b-V
Consumption of cadmium in the United States of America (in tons)

Year	1967	1968	1969	1970	1971	1972
Consumption	5,252	6,045	6,832	4,111	4,913	5,391

Table 3.2.b-VI
Cadmium consumption and production (in tons), in the United States of America and in five member states of the European Economic Community during 1970 and 1971

Country	Metallurgical production	Imports	Exports	Consumption
B/L	1,093 / *947*	719 *(1)* / *868 (1)*	828 / *735*	650 (s.i) / *703* (s.i)
F	528 / *579*	591 / *479*	91 / *89*	1,028 (s.e) / *950* (s.e)
UK	318 / *262*	1,171 / *858*	19 / *143*	1,313 (s.i) / *1,176* (s.i)
RFA	1,035 / *982*	1,224 / *962*	249 / *287*	2,010 (s.i) / *1,783* (s.i)
I	425 / *350*	31 / *20*	72 / *62*	370 (m) / *320* (m)
USA	4,295 / *3,579*	1,131 (metal) 504 (p) / *1,582* (m) 525 (p)	170 / *30*	5,260 (s.i) / *5,130*

(1) *Raw cadmium, waste material included but worked cadmium excluded*
(s.i) *Variation of stocks included*
(s.e) *Variation of stocks excluded*
(m) *Average consumption*
(p) *Powder*

During this period, the increase in consumption in Europe was slightly greater than 22%, while in the United States consumption remained stationary with, however, a peak production in 1969 and a large recession in 1970.

Considering the decade 1962-1972, the total increase in consumption of cadmium in Europe was about 50%, this being a larger increase than that of zinc (more rapid until 1969 and less rapid afterwards).

During the last decade, it appears that the efficiency of recovery of cadmium from zinc minerals has increased considerably. Nevertheless from 1970 the cadmium market has experienced some difficulties, probably due to problems associated with the protection of the environment.

Concerning secondary cadmium, recycled cadmium, (for example, recovery of cadmium from batteries), this apparently represented up to 20% of the metallurgical production of cadmium in the United States of America in 1969. Table 3.2.b-VI, taken from a private French reference source (57), shows, for five member states of the European Economic Community and for the United States, the consumption in tons of cadmium, based on the production, importation and exportation of each country.

These figures, compared to other statistics, specially those given in a study by Roskill Information Services Ltd., (17) for 1970, show certain marked differences. It is evident that each entry and departure of material across national borders will be recorded as imports or exports in the statistics of each country and will thus cause a distortion of the real consumption when all the countries are taken as a whole.

3.3 UTILIZATION

a. Zinc

a.1 Introduction

Five major types of industrial uses of zinc can be defined.

- **Use in the protection of surfaces against corrosion** (hot galvanizing, zinc paint, cathodic protection ...),
- **Use in alloys of zinc,**
- **Use in alloys of copper and light metals,**
- **Use of thin zinc sheets** in various forms (plates, leaves, bands) in the industrial and domestic sectors (covers, etc.),
- **Use of zinc oxide** in paints, glass and ceramics, chemical and pharmaceutical industries,

It should be recalled here that all these industrial applications lead to the production of objects either having surfaces of zinc (galvanized objects such as gutters, buckets or paints etc.), or with zinc included in them (batteries, glassware, ceramics), which are needed for domestic purposes. These objects are usually thrown away after use and thus become an important source of waste zinc in the environment.

Table 2.a in **Annexe 2**, at the end of this report, summarizes the various uses of zinc.

a.2 Divisions

Table 3.3.a-I below shows the principal uses of zinc, based on figures for the consumption of zinc slabs in 1970, in certain member states of the European Economic Community, and in the United States of America. To these figures have been added the tonnages of zinc coming from various recovery operations (92,000 t from zinc waste in the United Kingdom, 44,000 t from various recovery processes in Italy, 17,000 t from recasting operations in the Federal Republic of Germany).

Table 3.3.a-I
Division of % tonnage of zinc consumed in several member states of the European Economic Community and in the United States of America

Principal Uses	RFA	F	UK	I	USA
Galvanizing	34	32	26	36	38
Sheet products	17	30	7	6	3
Copper alloys	24	13	29	38	12
Light alloys	21	16	20	18	42
Zinc oxide (ZnO)	2	4	10	-	3
Various	2	5	8	2	2

Lastly, table 3.3.a-II gives, based on statistics for 1970, the % principal divisions of applications in the nine member states of the European Economic Community.

Table 3.3.a-II

Uses	Percentage
Galvanization (plate, tube, sections, wires, etc.)	32
Rolled and wire drawn zinc	15
Copper alloys	26
Light alloys	19
Various (ZnO, zinc powder, various salts, etc.)	8
TOTAL	100

a.3 Some observations on the zinc market

The zinc market is complex.

For example, zinc oxide (ZnO) produced by the oxidation of zinc metal only represents a certain fraction of raw material associated with the formation of zinc oxide and of zinc salts used in the chemical industry. Thus, the direct production of ZnO, in France, represents only 20% of the total production of ZnO.

In the Federal Republic of Germany, lithopone, of which about 50,000 tons was produced in 1970, an important part was obtained from zinc-pyrites, which is not considered as a zinc ore.

Equally, zinc powder may be produced from primary zinc but also by a reduction process.

a.4 Tendency of usage

For France in particular, the division of deliveries to industrial users has not fundamentally changed between 1970 and 1972, although galvanizing uses have slightly increased at the expense of alloys.

It seems, however, generally speaking, that galvanizing is the most important industrial use of zinc in most countries. In the United States of America for example, much research is carried out in an effort to increase its use in pre-fabricated materials used in the construction industry or to produce a better finish on galvanized objects (chrome polish on galvanized steels).

In the chemical industry, new organo-zinc compounds are now in use.

Zinc oxide is used as a stabilizer for U.V. in thermoplastic products.

Lastly, it should be noted that there is a diminishing tendency to use lithopone paints during recent years.

b. Cadmium

b.1 Introduction

Cadmium is used in many industrial sectors, which are summarized below:

- Cadmium plating,
- Production of alloys of low melting points,
- Production of batteries,
- Production of pigments for paints and ceramics,
- etc.

Table 2.b in **Annex 2** summarizes a certain number of uses of cadmium as well as the elements and salts employed.

b.2 Consumption

Table 3.3.b gives for 1970, the division in percentages, of the consumption of cadmium for two member states of the European Economic Community and for the United States of America.

Table 3.3.b (17), (53)

Uses \ Country	RFA	UK	USA
Electroplating	37	37	50
Pigments and stabilisers	38	38	40 (20+20)
Alloys and solders	10	10	7.5
Batteries	14	14	-
Various	1		

The figures given for Germany are taken from a report by Kienbaum (53) and those for the United Kingdom are taken from a report by Roskill (17). These figures are strictly similar and leave little hope as to their usefulness. The exact figures, including those for France, would be of great interest.

b.3 General observations on the reutilization of cadmium

1) **For cadmium plating**, the main losses are those that occur in the industrial surfaces treatment. They will be discussed in detail in the chapter on cyanides.

2) **Pigments**

- for paints, the losses are very scattered,
- for inks, cadmium can vaporize when burning, particularly newspapers,
- for stabilizers, the losses are equally scattered and recovery will be very difficult.

3) **Alloys and welding**

It is not intended to discuss this problem which was the theme of a colloquium by the Commission of European Communities, held in Luxembourg in 1973.

4) **Cadmium** may be easily recovered by volatilization, but, however, attention must be paid to the type of furnace employed.

5) It should be remembered that there is an industry sector which is concerned with the recovery of cadmium from various forms of industrial waste (scrap).

3.4 A PRELIMINARY DISCUSSION ON THE QUALITATIVE AND QUANTITATIVE WASTE PRODUCTS PRODUCED BY VARIOUS INDUSTRIES

a. Zinc

a.1 In production

Without rediscussing the two major extraction processes for zinc (thermal and wet production) mentioned briefly in paragraph 3.1.b above, it should be noted that in 1970 about **55%** of the **zinc produced** came from **electrolytic processes** (wet production) and the remaining **45%** from **thermal processes.**

In the future, two processes seem to be particularly important and these will be examined later, especially from the point of view of ascertaining possible losses to the environment:

- **blast furnace process or Imperial smelting process (thermal production),**

- **the Jarosite process (wet production).**

The two systems of treatment will be discussed separately except for "roasting" which is common to both processes.

1) Roasting

Taking the most modern system of roasting, the losses that occur with a **fluidized bed** (as used by Kokkola, Finland), 50 tons per day of ZnS, zinc blend, (50% of the concentrate is <50μ) are roasted with an air flow of 50,000 m^3 per day.

The quantity of dust remaining behind the "Cottrell" is normally about 120 mg/m^3, which is equivalent to a small loss of 120 kg/day. However, the waste gases contain between 50 to 100 ppm of mercury which is washed by concentrated sulphuric acid. After washing, there is some 0.1 to 0.2 ppm of mercury remaining on the vanadium catalyser which oxidises SO_2 to SO_3.

In the modern roasting plants, the only atmospheric pollution should be limited to residual SO_2 at the outlet (chimney) of sulphuric acid section.

2) Blast furnace process or Imperial Smelting process

Factories using this process are usually large and able to use mixed zinc and lead minerals, (for example 120,000 tons per year of zinc and 40,000 tons per year of lead).

Without going into detail, the outline of the three main processes is given below:

. **Agglomeration** (sinterisation) of calcines and of dusts, recycling in order to form the sinter mix of coke and chalk.

. **Reduction** by the following endothermic reactions:

$$ZnO + CO \rightarrow Zn + CO_2 - q$$

$$CO_2 + C \rightarrow 2CO - q'$$

Air, entering the base of the blast furnace allows the coke to burn, thus producing the heat (calories) necessary for the above reaction to take place (greater than 950°C at the mouth of the furnace).

Zinc vapours which condense quickly are collected in the upper part of the blast furnace, work lead is recovered with the slag in the bottom of the shaft.

. **Purification** (refining) which is carried out by eliquation or eventually by distillation.

The **purity of zinc** produced by this method is about **92%** and thus gives an initial idea of the losses (about 6%) in the slag which is found on the surface of the work lead and in the dust formed in the course of the mineral conditioning and sintering operations. Some losses occur as well in the condensation and purification processes.

The slag is mostly composed of non-reduced zinc, iron and silicon oxides and represents about 30% of the weight of the initial charge of minerals.

An example of the approximate composition of the residue is given below:

. Zn	9 %
. Pb	1 %
. Cu	0.5 %
. Cd	traces
. Fe	30 %
. As	0.2 %

3) The Jarosite electrolytic process

The most important phases of this process, which occur in solution, are briefly mentioned below:

. **Leaching** in two steps of the roasted minerals:

- attack at neutral pH,

- attack at acid pH, normally the solution is heated in order to dissolve zinc ferrites contained in the material coming from the first stage.

. **Purification** of solutions

- the solution of neutral pH is purified through sedimentation (with the addition of zinc powder) and then goes to the electrolysis cells;

- the solution of acid pH is filtered and then precipitated by addition of NH_3 and ammonium sulphate.

. Iron and ammonium sulphate (**Jarosite**) precipitate, and the filtered solution is returned to step one (attack of neutral pH).

The separation of iron is selective in the presence of alkaline, and is the same as the precipitation of a basic sulphate $X_2\ (Fe_6(SO_4)_4(OH)_{12})$

where X is Na, K or NH_4

. **Electrolysis** of zinc sulphate ($ZnSO_4$) is according to the reaction:

$$ZnSO_4 + H_2O + (e) \rightarrow Zn + H_2SO_4 + \tfrac{1}{2}O_2$$

The acid formed in the electrolytic baths is returned to step two (attack at acid pH).

Concerning the **losses of zinc**, two important residues are recovered and treated.

. **The residue from filtration after acid attack**, having a weight of about 10% of the initial mineral, contains between 7 and 8% of zinc.

A typical composition of this residue is given below:

. Zn	7 to 8 %
. Cd	0.15 %
. Cu	0.10 %
. As	10 %

. **The residue from the precipitate Jarosite** having a weight of about 30% of the roasted mineral contains between 5 and 6% zinc.

It is worth remembering the composition of these residues, difficult to be disposed of, due to the amount produced:

. Zn	5 to 6 %
. Fe	30 %
. Cu	0.3 %
. Cd	0.04 %
. As	0.4 %

The loss due to these precipitates represents about 2.5% of the initial weight of zinc.

Finally, there are also some losses in the cementation muds; **the total yield of zinc** by electrolytic processes is thus estimated to be about **95%**.

a.2 Utilization

1) Introduction

The industries that are most likely to discharge large quantities of zinc are the following:

- steel works using galvanizing processes,

- manufacturers making zinc containing materials such as bronze and brass,

- manufacturers using electroplating processes of zinc and bronze,

- manufacturers of silver plating,

- rayon viscose production,

- production of papers and magazines etc.

Without discussing in detail all the various types of wastes, a brief outline of the effluents produced by galvanizing and electroplating processes, as well as those in the production of viscose, is given as an example of their type.

2) Galvanizing and electroplating processes

Approximately one third of the zinc consumed in the nine member states of the European Economic Community is used in galvanizing processes. The main sources of loss come from fluxing operations (the flux being a mixture of zinc chloride and ammonium chloride) and sweeping operations regulating thickness of plating.

For zinc electroplating, the losses are of a similar order to other forms of electroplating.

The zinc solutions used in electroplating contain between 5 to 35 g/l zinc; they are replenished from time to time. The concentration of zinc in rinsing waters is a function of the bath concentration, the draining, and the volumes used.

The baths and the rinsing waters, generally, also contain cyanides. This point will be discussed in Chapter 4.

3) Production of viscose rayon (rayon)

In the production of rayon textiles and threads, zinc sulphate is added to the production baths in order to control the coagulation (solidification) of the thread. A certain amount of $ZnSO_4$ is caught by the thread as it leaves the bath, and is found in the rinsing waters.

Sometimes, the first rinsing bath (the most concentrated) is concentrated by evaporation and $ZnSO_4$, and other salts present are reused in the preparation of new baths. On the other hand, rinsing waters which are less concentrated are invariably mixed with the factory's waste effluents.

In the United States of America, the production of viscose rayon amounts to about 500,000 tons annually, consuming approximately 10,000 tons of zinc which are practically all lost as waste.

Based on figures of consumption of zinc, in industries producing viscose rayon, in the European Economic Community and EFTA, it is possible to estimate the annual losses of zinc as 10,000 tons, for a total Western European production of some 900,000 tons of rayon.

4) Other sectors

Other sources of wastes containing zinc may be mentioned:

- mine waters,

- paper manufacture (bleaching with zinc dithionite),

- fungicides (zinc diethyl dithio carbamate),

- paint production (zinc tetrahydroxy chromate) and pigments (lithopone - ZnS, $BaSO_4$).

Finally, it should be noted that the use of zinc in the production of dry batteries could be a method for the dispersion of zinc containing residues in the environment.

b. Cadmium

b.1 In production

The main sources of supply in the production of zinc and cadmium have already been mentioned (see figures 3.1.b-I, II and III).

Cadmium can, in fact, be obtained by four main routes:

- production of zinc by thermal or electro-thermal processes,

- production of zinc by electrolylitical processes,

- copper and lead foundries.

The percentage of cadmium produced throughout the world from lead metallurgical processes probably does not exceed 5 to 6%, the major source of cadmium being from zinc production.

Table 3.4.b-I gives the percentage of cadmium, produced worldwide, in relation to the production of zinc.

Table 3.4.b-I
Production of cadmium and zinc, in tons, in the world from 1969 to 1972 (17)

Year \ World production	Cadmium	Zinc	Percentage
1969	17,576	5,170,000	0.33
1970	16,544	5,102,000	0.32
1971	15,501	5,004,000	0.31
1972	16,740	5,422,000	0.31

1) **Electrolytic process** (see figure 3.1.b-II)

In this procedure, one of the most widely used, cadmium is recovered in the purification step as the solutions go for electrolysis, by sedimentation of the zinc dust. The precipitate obtained is then dissolved (alkaline wash), and afterwards electrolysed in cells having aluminium cathodes and insoluble lead anodes containing 1% silver. The cadmium recovered at the cathodes is re-dissolved, the final product having a purity of the order of 99,95%.

Cadmium losses occur mainly in residual waters coming from purging operations of the cells of electrolysis and may be in the order of a few percent (10% maximum) of the amount of cadmium present in the initial load of minerals.

2) **Thermal processes** (see figure 3.1.b-III)

In these processes, cadmium is found almost completely in the smoke and dust coming from roasting and sintering operations. As well as cadmium other elements are found such as zinc, lead, arsenic etc. The solids collected after "Cottrell" and condensation operations in the form of oxides, are washed in acid solutions and the cadmium precipitated, after which it is purified by fusing with caustic soda to eliminate zinc and other impurities.

Other than those losses due to purges of the system, various other operations result in losses as vapour or smoke which escape to the atmosphere, especially when no recovery systems are in place such as the "Cottrell" type.

In this last case, the losses of cadmium to the environment may reach several percent of the quantity of metal in the initial load of minerals.

It should be noted, however, that the solid materials recovered by a "Cottrell" type unit are not always recycled and are sometimes dumped on waste-heaps if the concentration of the element required is not sufficiently high to justify re-treatment. In this case, waste-heaps may become a source of pollution because of natural processes such as wind dispersion and natural leaching.

3) Other sources of cadmium

Other secondary sources of cadmium that should be mentioned are:

- cadmium residues from the pigment industry,

- recovery from used batteries.

The amount of cadmium, produced throughout the world by these methods is probably not greater than 4 to 5% of the world production of this metal.

Due to its physical properties, low melting (321°C) and boiling (767°C) points, great care must be taken in the operation of ovens for the "distillation" of used batteries.

b.2 In utilization

1) Introduction

The main industrial uses of cadmium are given below:

- electroplating,

- production of pigments and stabilizers,

- production of alloys (of low melting point - used in production of taps),

- production of stabilizers for p.v.c.,

- production of batteries,

- production of television tubes,

- catylists,

- fungicides,

- photographic industry (light meters) and solar cells,

- rubber industry,

- production of "central rods" in the nuclear power industry,

- de-oxidizing agents.

To supplement table 3.4.b-I given above, table 3.4.b-II gives the percentage of cadmium used in various processes, for the United Kingdom, the United States of America and Japan. The figures given below are taken from an English reference source (17) and are for the year 1972.

Table 3.4.b-II
Main uses of cadmium in the United Kingdom, United States of America and Japan in 1972 (percentage of cadmium used)

Uses \ Country	UK	USA	JAPAN
Electroplating	35	55-60	10-15
Pigments	41.6	10-15	25
P.v.c. stabilizers	-	15-20	20
Batteries	-	3-5	10
Alloys	6.3	-	10
Various	16.5 (includes batteries)	5-10	20-25

Although figures are not available for European countries other than the United Kingdom, the various percentages are most likely to be similar for the other member states of the European Economic Community.

Table 3.4.b-III shows the main changes in utilization of cadmium in the United Kingdom for the years 1970 to 1973.

Table 3.4.b-III
Principal uses, in tons, of cadmium in the United Kingdom

Uses \ Year	1970	1971	1972	1973
Electroplating anodes	370.9 } 490.4	314.0 } 421.9	306.8 } 478.7	292.3 } 468.8
Electroplating salts	119.5	107.9	171.9	176.5
Copper alloys	33.7	35.4	32.6	28.6
Other alloys	15.3	14.0	15.0	17.5
Welding	84.0	46.4	38.0	38.6
Pigments	502.5	461.0	566.0	660.0
Various (batteries included)	186.9	197.5	223.7	328.8
Total Consumption (in tons)	1,321.8	1,176.2	1,354.0	1,542.3

2) Losses associated with the major uses of cadmium

Little information exists for the various member states of the European Economic Community concerning the precise tonnage of cadmium consumed in the principal industrial sectors, such as electroplating, pigments and various uses including the production of batteries.

However, on the basis of available figures (see table 3.2.b-IV and 3.4.b-III), as well as the model proposed in paragraph b.2.1 above, table 3.4.b-IV below attempts to evaluate the losses of cadmium in wastes, produced by its major uses by member states of the European Economic Community.

Table 3.4.b-IV
An attempt at evaluating the overall losses of cadmium in wastes, produced by its major uses, in the nine member states of the European Economic Community (based on the figures for 1972)

Input (in tons) and % of total consumption	Principal uses	Output losses in waste in tons
2050 (35%)	Electroplating (cadmium)	1,000 to 1,500 ≃ 50 to 75%
2470 (42%)	Pigments	125 to 250 ≃ 5 10%
970 (16,5%)	Various (including batteries) *	?

It must be noted that bath and rinsing water operations in electroplating processes generally contain cyanides, and these will be discussed in Chapter 4.

3) Other uses

Other sources of potential cadmium pollution exist and require careful examination, in particular:

- certain plastics containing stabilizers of cadmium/ barium (barium stearate),

- certain low melting point alloys which are used in plumbing and sanitary operations (this point requires special consideration because of the real dangers of contamination of fresh water drinking supplies),

- fungicides and insecticides (mite repellant) of the type cadmium pentachlorophenate, cadmium anthranilate etc.,

- photographic industry (cadmium bromide, iodide etc.),

- porcelain and glass production (cadmium nitrate),

- photo-electric sensitized cellulose.

**In general, cadmium is recovered from spent batteries with very high efficiency (≃ 100%)*

3.5 SUMMARY OF THE PREVENTATIVE TECHNIQUES KNOWN FOR EFFLUENTS CONTAINING ZINC AND/OR CADMIUM

a. Zinc

a.1 General

In chapter 3.4 above, an examination of the two major types of industrial effluents which may be sources of zinc pollution in the environment, have been carried out. These are respectively effluents coming from primary production of zinc and effluents resulting from industrial and domestic uses of zinc.

Effluents coming from various production operations may contain, as well as zinc, many other elements such as mercury, the principal waste product of grilling, arsenic, cadmium etc., which mainly result from purification processes.

Table 3.5.a-I below, shows the concentration of zinc in the principal effluents of various industries.

Table 3.5.a-I
Principal sources of industrial effluents containing zinc concentrations

Source	Concentration of zinc in mg/l of effluent range found
Metallurgy polishing	0.2 - 1,500
Galvanizing Electroplating	2 - 350
Silverplating	0 - 220
Production of rayon	20 - 1,000

Various treatment techniques are used in order to reduce, recover or eliminate zinc in industrial effluents:

- the technique of chemical precipitation, this may take place in one or two successive precipitations,

- ion exchange techniques.

Other methods such as **electrolysis**, combined with the passage through an ion exchange column, as well as by concentration seem also to show promise but require a lot more basic research to be carried out.

a.2 Simple precipitation procedure

This process, often used in order to eliminate various metallic elements in waste waters, consists essentially of adjusting the pH of the solution by the addition of chalk or soda (NaOH, Na_2CO_3, $Ca(OH)_2$) so that a precipitate of zinc hydroxide is obtained*.

With sodium sulphate by adding CaO, we get a simultaneous precipitation of $Zn(OH)_2$ and $CaSO_4$.

In order to eliminate zinc, one can precipitate in the sulphide form, using either a sulphide, an alkaline sulphate, alkaline earth or even possibly hydrogen sulphide.

One of the principal disadvantages of this process is that the inorganic muds (sludge) obtained must be discharged into specially designed ponds or reservoirs in order to prevent contamination of fresh waters.

Also this process may be difficult to incorporate into a recycling system, which in the long term makes it uneconomical.

a.3 Double precipitation procedure

In this procedure, a concentrated solution of zinc hydroxide (the zinc concentration being at least 50 times greater than that in the effluent), is added to the effluent and is treated in a special floculation tank. The recycled hydroxide reacts with the zinc in the effluent and is precipitated. After the first contact, ammonia solution is added to complete the reaction and to maintain the pH at between 9 and 10. The addition of a coagulant is not required because of the great density of the mixture.

In the case of a rayon producing factory, the effluents, which may well contain 200 mg/l zinc, are treated with caustic soda in a floculation tank, and it is possible to obtain a concentration of 5% zinc hydroxide by simple decantation.

If, when the same effluent is treated with caustic soda at pH $\simeq$ 9 without passing through the floculation tank, the concentration of zinc hydroxide is less than 1% and the precipitate is very difficult to handle.

**With good adjustment of pH followed by filtration on pure sand, it is quite possible to lower the zinc concentration below a mg/l (p.p.m.)*

Before passing through the floculating tank, certain impurities must be eliminated from the effluent. This may be easily carried out by treating the acid effluent with chalk or caustic soda at a pH between 5 and 6.

After the second stage of the precipitation, the solids are decanted, collected in a tank and gently agitated before being recycled to the head of the circuit or finally eliminated from the floculator. A further decantation thickens the precipitate still further until the concentration of hydroxide reaches 5% as mentioned above.

After a series of experiments on pilot plants, this procedure has been successfully used on effluent from an American factory* producing rayon, an affiliated member of the European Oceanic Association.

It is interesting to note that this particular procedure can be used on a very large scale, not only for the recycling of zinc but also for the separation of zinc and magnesium. Magnesium that is recovered may be re-used in production processes.

a.4 Ion exchange techniques

Much information exists on the purification techniques using ion exchange on industrial waste waters. These techniques may be used for recovery, or for purification of effluent residuals.

Unless zinc is associated with other rare elements, or expensive to treat or to replace, ion exchange techniques seem, in general, to be difficult to justify on economic grounds.

In the rayon producing industry, a procedure has been put into operation which allows recycling of zinc. This procedure, which uses non-selective acid resins, of the type sulphonated acid styrenes may be used when the concentration of calcium is very low.

However, a new zinc selective resin has just been brought into service and is at present under intensive study for use in the treatment of effluents containing high concentrations of calcium.

A combination of these two resins seems a promising solution for the treatment of effluents produced by the rayon industry, and will circumvent the necessity for all precipitation procedures.

There is however a great deal of research still required in this area (new resins, influence of various impurities etc.), before an accurate economic comparison with other methods can be made.

**American Enka Corporation*

a.5 Electrolysis combined with the use of ion exchange resins

In spite of the well known advantages of electrolysis for the recovery of metallic elements from solution in various types of industrial effluents, this method has practically never been employed up to now because of economic considerations.

Nevertheless, recent research on a two step method, comprising an electrolysis step followed by purification on ion exchange columns, has shown its usefulness, both from a technical as well as economic view point. These preliminary results have now to be verified using industrial pilot plant units.

a.6 Treatment using evaporation techniques

In this technique, the effluents are concentrated by evaporation and then undergo several counter-current washes.

a.7 General economic considerations on the main preventative techniques summarized above

The figures given below correspond to an industrial treatment plant having an effluent capacity of 600 to 700 m^3/h coming from the production of rayon, i.e. having 60 kg/h of zinc.

Table 3.5.a-II gives the investment costs of three different treatment installations as well as the yearly running costs.

Table 3.5.a-II *(private source)*

Cost in thousands of units of account u.c.	Simple precipitation *(cf § 3.5.a.2)*	Double precipitation *(cf § 3.5.a.3)*	Ion exchange with recycling *(cf § 3.5.a.4)*
Investment costs	340	544	852
Running costs	340	442	680
Total	680	986	1,532
Recycled products (value in thousands of u.c.)	-	479	479
Total costs	680	507	1,053

On the basis of the recovery of 500 tons of zinc per year, table 3.5.a-III below gives the costs of zinc recovered by the three possible processes.

Table 3.5.a-III

Process employed	Cost of recovered zinc
Simple precipitation *(cf § 3.5.a.2)*	≃ 1.38 u.c/kg
Double precipitation *(cf § 3.5.a.3)*	≃ 0.92 u.c/kg
Ion exchange with recycling *(cf § 3.5.a.4)*	≃ 2.12 u.c/kg

Thus the cost of treatment by ion exchange and recycling would be double the cost by precipitation and recycling, as has already been mentioned in the American literature (49).

b. Cadmium

b.1 General

Table 3.5.b-I gives the concentration of cadmium in various kinds of effluents discharged by industry (49).

Table 3.5.b-I
Main sources of industrial effluents containing cadmium and ranges of concentration

Sources	Concentration of cadmium in mg/l in effluents
Washing baths in plating processes	7 to 50
Electroplating baths	23,000
Passivation baths and electro polishing	2,000 - 5,000

Concerning the prevention of cadmium in industrial effluents three main types of treatment may be considered:

- chemical precipitation technique,

- ion exchange technique,

- concentration by evaporation techniques as well as other methods.

b.2 Chemical precipitation technique

By adjusting the pH (alkaline), cadmium may be precipitated in the hydroxide form, the precipitate obtained being insoluble and very stable.

For pH values near 8, the method reduces the concentration of residual cadmium to 1 mg/l. This value may be lowered to 0.1 mg/l with pH values near 10.

If iron is present, cadmium coprecipitates with iron hydroxide at pH 8.5 and the residual cadmium concentration is lowered to 0.05 mg/l. If sufficient quantities of hydrated chalk are used, it is possible to lower the residual cadmium to 0.01 mg/l in the effluents so treated.

As for zinc, it would be possible to precipitate cadmium in the form of CdS which is particularly insoluble.

Lastly it should be noted that the effect of complexing agents, such as CN for example, inhibits the precipitation of cadmium and it is thus necessary to destroy them by pretreatment of the wastes.

b.3 Ion exchange techniques

In principle, treatment of effluents by ion exchange is used as a method of recovery or purification of cadmium so that recycling of waters is possible.

Effluents concentrated by passage through ion exchange columns will be, in general, more easily treated for economic recovery of cadmium than those effluents only slightly concentrated, coming from various factories using this element.

At the present time, treatment by ion exchange largely remains a laboratory technique and only a few pilot plant units, using both anionic and cationic resins, are actually in the process of being studied and developed.

b.4 Concentration of effluents by evaporation techniques and other methods

Depending upon the variation of the price of cadmium*, its recovery from various industrial effluents changes greatly:

- **recovery by highly sophisticated processes such as inverse osmosis, electrodialysis, etc.,** require the metal to be expensive in order to make it economic,

- **recovery by evaporation techniques** is dependent on the high cost of energy.

**1970 46 F/kg*
1972 25 F/kg
1974 45 F/kg

CHAPTER 4

ORIGINS, FORMS AND CHARACTERISTICS OF POLLUTION BY CYANO-METAL COMPOUNDS

CHAPTER 4 - ORIGINS, FORMS AND CHARACTERISTICS OF POLLUTION BY CYANO-METAL COMPOUNDS

4.1 SUMMARY OF LARGE-SCALE METHODS OF PRODUCTION OF CYANIDES AND CYANO-METAL COMPOUNDS

a. Hydrocyanic acid - HCN

a.1 Methane procedure

The most widely commercially used method in the last ten years is the methane procedure. It depends upon the reaction of methane (CH_4) with ammonia (NH_3) in a current of air in the presence of a platinum catalyst. The reaction takes place at a temperature of between 1,000 and 1,200°C, according to the equation:

$$CH_4 + NH_3 + \frac{3}{2} O_2 \rightarrow HCN + 3H_2O$$

In this procedure (Andrussow method), the exhaust gas must be rapidly cooled. The hydrocyanic acid is diluted and it is necessary to recycle any unreacted ammonia.

Several variations (Degussa ...) allow a higher concentration of hydrocyanic acid at the end of the reaction or even the direct formation of the salt sodium cyanide NaCN. It is also possible to use propane (C_3H_8) instead of methane.

a.2 More and more often these days the principal production of hydrocyanic acid is from the formation of acrylonitrite (starting from propylene and ammonia) or by the formation of methyl-metacrylate, in which HCN is present as a bi-product.

The recuperation of HCN by scrubbing of the exhaust from the coke furnace has not been considered a satisfactory procedure for the last ten years.

a.3 The production of hydrocyanic acid from sodium cyanide and mineral acids will also be discussed.

Finally, in Germany, the formamide process is still in use although on the decrease. The formamide is produced in the reaction of carbon monoxide with ammonia which then decomposes to HCN and H_2O according to the reaction:

$$HCONH_2 \rightarrow HCN + H_2O$$

b. **Alkaline cyanides**

b.1 Castner process

In this procedure, metallic sodium reacts with charcoal and ammonia according to the reaction:

$$2Na + C + 2NH_3 \rightarrow 2NaCN + 3H_2$$

This reaction occurs at a temperature of 800°C and has intrinsic dangers for the industrial plant and is thus decreasingly employed.

b.2 The Castner process is being replaced by a method consisting of the neutralization of HCN by sodium hydroxide solution (or potash) to form alkaline cyanides such as NaCN or KCN. However, it should be underlined that the method is also not without its dangers.

c. **Calcium cyanide**

A relatively important quantity of $Ca(CN)_2$ is still produced (for the formation of precious metal cyanides) by heating, in an electric furnace to 1,000°C, unpurified calcium cyanide (containing a small quantity of carbon) in the presence of sodium chloride:

$$CN_2Ca + C \rightarrow Ca(CN)_2$$

The melted product is cooled in order to stop re-decomposition to CN_2Ca.

d. **Other cyano-metallic compounds**

Other cyano-metallic compounds may also be obtained by the neutralization of HCN or by precipitation or double decomposition of alkaline cyanides.

4.2 MAIN INDUSTRIAL SECTORS CONCERNED IN THE USE OF CYANIDES AND CYANO-METALLIC COMPOUNDS

Many industrial sectors are concerned with cyanides and cyano-compounds, either in the form of metal compounds used in production or in their direct consumption or as bi-production in various industrial processes.

A summary of the various industrial sectors and main methods of production is given below, as outlined in Chapter 4.1 above.

a. **Electroplating**

This sector is one of the most important as a source of cyanide waste products and thus requires much attention, especially in view of the numerous small electroplating workshops in existence, where the control of effluents is of primary concern. In this type of industry large quantities of electrolytes, containing cyanide, are employed.

Although, for some years, new families of electrolytic complexing agents have been used (fluoborates, fluosilicates, sulfamates, pyrophosphates, gluconates etc.), cyanides are still largely in use, thus requiring treatment of the used concentrated solution. Waste solutions occur from different stages of rinsing, as well as other processes such as:

- rinsing water from reaction vessels, filters and soils; a typical composition is given below:

  . Cu : 50-60 mg/l ; Zn : 20-25 mg/l ; Cd : 15-20 mg/l

  . CN : 80-100 mg/l ; free alkalines : 80-100 mg/l ; pH ≃ 11,

  It should be noted that waters containing alkaline also contain chromium at varying acidities.

- aerosols resulting from washing processes, etc.

Concerning more specially electroplating operations which involve metals under study in the present report, only processes of **cadmium** and **zinc plating** use cyano-electrolytes and produce effluents which contain salts of cadmium (CdCN) and zinc cyanide (ZnCN).

It should be mentioned that it is also possible to find in waste solutions produced in electroplating, quantities of sulfocyanide (SCN).

In general, chromium electroplating does not use cyano-compounds in the electrolytic solution employed in this process.

Table 4.2. below, gives the composition of various electrolytes used in cadmium and zinc electroplating.

b. **Mining industry**

In the treatment of certain minerals and specially in the recovery of silver and gold, sodium cyanide (NaCN) is used as transfer agent:

$$AgCl + 2NaCN \rightarrow Na(Ag(CN)_2) + NaCl$$

$$2Au + 4NaCN + O_2 + 2H_2O \rightarrow 2Na(Au(CN)_2) + 2NaOH + H_2O$$

TABLE 4.2.a - Composition of various electrolytes used in cadmium and zinc plating (44)

CONSTITUANTS in g/l	CADMIUM PLATING ELECTROLYTES		ZINC PLATING ELECTROLYTES						
			ACID BATHS		FLUOBORATE		PYROPHOSPHATE	CYANIDES	
	"Mort"	"Tonneau"	Sulfate	Chloride "mort or tonneau"	"Mort"	"Tonneau"		"Mort"	"Tonneau"
Cadmium oxide	22.5	22.5							
Zinc oxide								42 to 55	
Zinc cyanide									60
Zinc pyrophosphate							50 to 70		
Zinc sulfate			180						
Zinc chloride			14	100 - 160					
Zinc fluoborate					200	136			
Sodium cyanide									
Sodium pyrophosphate							190 to 200		
Sodium cyanide	78	85.8						90 - 105	30
Caustic soda		7.5						35 - 100	45
Ammonium citrate							20 to 30		
Ammonium chloride					27	54			
Ammonium Fluoborate					35	35			
Sodium silicate							0.5 to 1.0		
Thio-urea							0.1 to 0.15		
Glue or liquorice				0.5 - 10	1	1	1		
Polishing	+	+	+	+	+	+	+	+	+
Nickel sulfate	0.5	0.5							

c. Metallurgy

The metallurgical industries principally use cyanides in two ways:

- thermal treatment,
- cleaning processes.

d. Synthetic fibre industry

Important quantities of cyanides are used in the production of certain synthetic fibres, especially acrylic fibres, (orlon).

e. Other sectors

Among others:

- coke producing industry,
- photographic industry (film development),
- production of synthetic organic compounds (monomers),
- installation of gas scrubbers for blast furnaces of steel making industry, etc.

4.3 CENSUS OF VARIOUS KNOWN PREVENTION TECHNIQUES FOR EFFLUENTS CONTAINING CYANIDES AND CYANO-METAL COMPOUNDS

a. General

Because of the negligible toxicity of cyanates in comparison to cyanides, one of the major treatment techniques is to transform, by partial oxidation, cyanides contained in effluents to cyanates.

When an absolute cleansing treatment of the waste is required, the cyanates themselves are completely oxidized and are totally destroyed, forming nitrogen and carbon dioxide.

This oxidation may be performed in several ways:

- oxidation by chlorine,
- by various peroxides (Caro acid, ammonium-di-persulphate, hydrogen peroxide),
- electrolytic oxidation,
- catalystic oxidation,
- ozone oxidation, etc.

FIGURE 4.3.a-I - SCHEMATIC DIAGRAM OF CLASSIC TREATMENT BY TRANSFORMATION OF CYANIDES INTO CYANATES

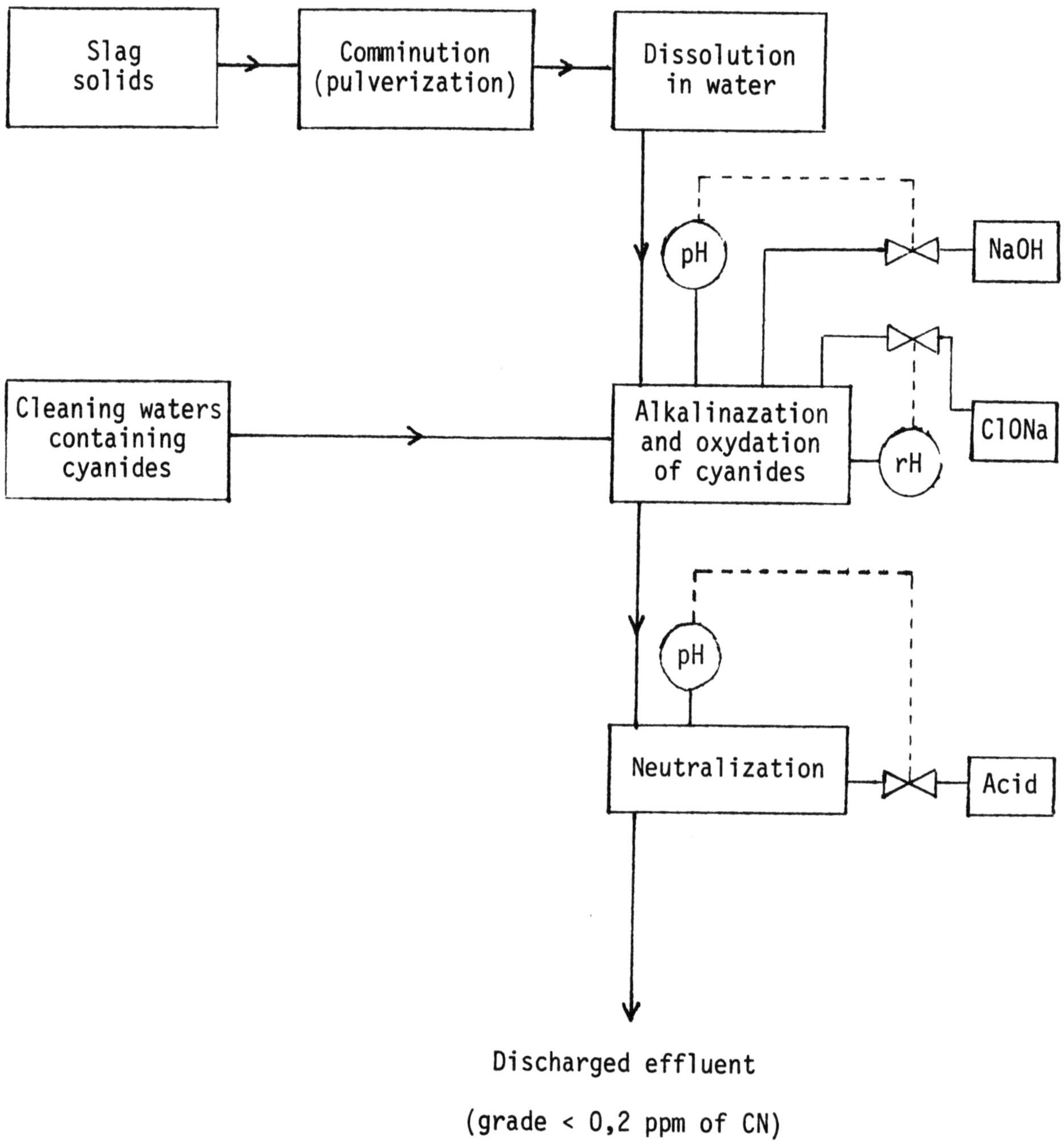

Various other complementary methods of treatment exist for cyanide containing effluents such as:

- evaporation method,
- ion exchange,
- inverse osmosis,
- active carbon method.

These may be applied to the various effluents summarized in Chapter 4.2 above.

Table 4.3.a-II gives some concentration ranges in cyanide effluents coming from electroplating operations (49).

Table 4.3.a-II

Sources of effluents	Concentration ranges (extreme) in cyanides mg/l
Rinsing waters	0.3 - 256
Glazing baths	15 - 20
Cadmium electroplating baths	20,000 - 67,000
Zinc electroplating baths	4,000 - 64,000
Silver electroplating baths	12,000 - 60,000
Copper electroplating baths	15,000 - 52,000
Bronze electroplating baths	40,000 - 50,000
Tin/Zinc electroplating baths	40,000 - 50,000

Table 4.3.a-III below, gives the typical concentration of effluents coming from a coke oven (52).

Table 4.3.a-III

Composition of effluents	Concentration mg/l
Ammonia	4,000
Phenol	2,235
Cyano Compounds	700
C.O.D. *(1)*	6,600
O.S.C. *(2)*	1,900
T.O.C. *(3)*	2,100
Suspended solids	50
pH	7.2

b. Destruction procedure by chlorine (hypochlorite or chlorine gas)

b.1 Sodium hypochlorite

The total destruction reaction is given below

$$CN^- + NaOCl \rightarrow CNO^- + NaCl$$

This takes place in two stages:

$$CN^- + ClO^- + H_2O \rightarrow CNCl + 2OH^-$$

$$CNCl + 2OH^- \rightarrow CNO^- + Cl^- + H_2O$$

On acidifying the solution obtained after the first reaction, it is possible to completely decompose the cyanates formed:

$$2CNO^- + 3ClO^- + 2H^+ \rightarrow 2CO_2 + N_2 + H_2O + 3Cl^-$$

Although in principal, chlorine gas can also be used, the relative high cost of storage and control installation for chlorine gas is not always justified (in the case of small installations) and thus hypochlorite is chosen as the oxidizing agent.

(1) Chemical oxygen demand
(2) Organic soluble carbon
(3) Total organic carbon

As well as hypochlorite, sodium hydroxide (NaOH) and hydrochloric acid (HCl) are used for adjustment in obtaining correct pH.

The amount of hypochlorite needed is a function of the oxidisable materials, organic or inorganic, present in the effluent.

The necessary equipment for this type of treatment is, generally, very simple (agitation vessels) and is combined, in principle, to precipitation equipment for the elimination of heavy metals present in the effluent. Control of hypochlorite and other reagents is made by classic electrical means (28) and (29).

b.2 Chlorine

In an alkaline medium, chlorine, coming from hypochlorite, causes destruction according to the following principal type reactions:

$$CN^- + 2NaOH + Cl_2 \rightarrow 2NaCl + CNO^- + H_2O$$

$$2CNO^- + 4NaOH + 3Cl_2 \rightarrow NaCl + 2Cl^- + 2CO_2 + N_2 + H_2O$$

c. Treatment with peroxides

c.1 Caro acid

Caro acid is a solution of monopersulphuric acid 200 g/l, active reagent H_2SO_5.

All simple cyanides or cyano-metal compounds such as CdCN, ZnCN etc., are rapidly oxidized in a few minutes at pH 9.

$$CN^- + SO_5^{=-} \rightarrow CNO^- + SO_4^{=-}$$

Sulfocyanide found in industrial surface treatment workshops (factories) or cokeries, is oxidized in about 10 minutes at pH 9 or 9.5 according to the reaction:

$$SCN^- + 4SO_5^{=-} + H_2O \rightarrow 5SO_4^{=-} + CNO^- + 2H^+$$

The concentration of sulfocyanide after treatment is always less than 0.1 mg/l.

Electrochemical characteristics of the reaction allow this method to be used in automatic treatment stations, functioning continuously. Effluents leaving these stations always contain less than 0.1 mg/l cyanides, except in the case where ferro or ferricyanides (not attacked) are present.

c.2 Ammonium persulphate

Although more economic than Caro acid, the reactions involving ammonium persulphate are much slower and this method is better used for the discontinuous treatment of wastes coming from surface treatment factories or from regeneration eluents from ion exchangers.

c.3 Hydrogen peroxide

The oxidation reaction is:

$$CN^- + H_2O_2 \rightarrow CNO^- + H_2O$$

This reaction must be carried out at pH 9 in order to stop the possible formation of very toxic hydro-cyanic acid.

The rate of reaction becomes interesting when the cyanide ion is associated with heavy metal cations, especially copper, used as a catalyst. This method is useful for purifying effluents from electroplating processes.

The method can be used either for the discontinuous treatment of concentrated cyanides from electroplating baths, or for continuous treatment of rinsing wastes (25).

d. Electrolytic oxidation

Cyanides are oxidized at the anode to CO_2 and N_2 with the production of CNO^- as an intermediary compound.

The free metallic ions or those liberated by dissociation of the cyano-metallic compounds are fixed at the cathode. This is a further useful point of this procedure especially for copper cyanides, as it eliminates metals thus making this an integral part of the purification process.

e. Catalytic oxidation

Semi-concentrated or concentrated cyanide waste is acidified to pH 2.5 and the hydrocyanic acid formed is liberated by air passing through a contact column having a Rachig ring. The gas mixture (air + HCN) is burnt in a furnace (maintained at low pressure) in the presence of a catalyser, and CO_2 and N_2 are formed. This process is called the Cyancat process. Figure 4.3.e below shows the installation scheme for automatic treatment of concentrated cyanide effluents by the Cyancat process (30).

f. Ozone oxidation

This method of purification is still in the laboratory research stage.

The oxidation reaction of cyanides to cyanates takes between 10 and 15 minutes at a pH between 9 and 12, it becomes quasi-instantaneous in the presence of a copper catalyst.

Oxidation by ozone is a possible alternative treatment to chlorine.

FIGURE 4.3.e - INSTALLATION SCHEME FOR AUTOMATIC TREATMENT OF CONCENTRATED CYANIDE EFFLUENTS BY THE "CYANCAT" PROCESS

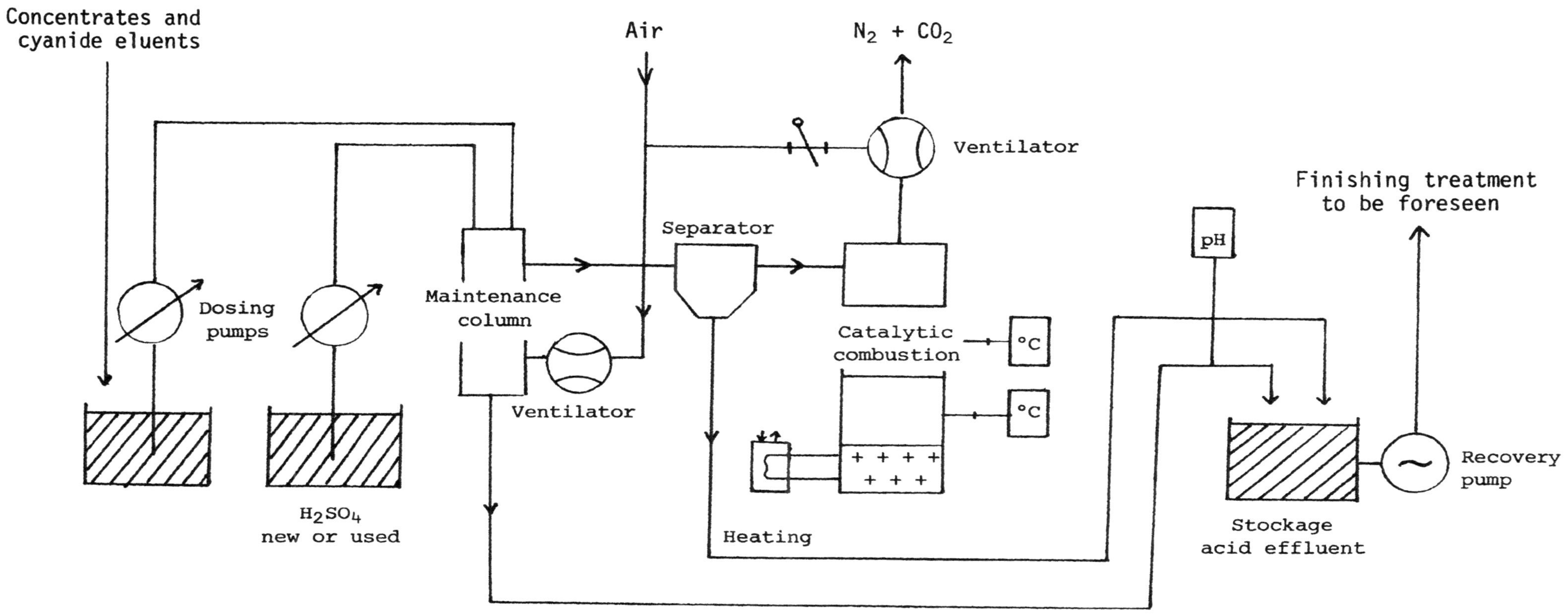

g. **E.I. Dupont process (Nemours)**

The Dupont Co. (Nemours), has recently patented a process of purification using hydrogen peroxide and formaldehyde in order to treat cyanide wastes, at a temperature of approximately 50°C.

The decomposition products comprise cyanates, ammonia and organic acids. In this procedure heavy metals such as zinc and cadmium are simultaneously precipitated if contained in the wastes.

h. **Other complementary methods of treatment**

h.1 Evaporation method

This method would be more particularly adapted to the treatment of rinsing waters from electroplating operations and could work under closed circuit, the concentrated solution returning to the electroplating baths and the water to rinsing reservoirs.

h.2 Ion exchange

This method is only mentioned in passing as it is now hardly ever used for the treatment of cyanide containing effluents. This is due partly to the deterioration of resins by the cyanide and partly to the danger of formation of hydrocyanic acid.

h.3 Inverse osmosis

The method most frequently used for the treatment of effluents containing cyanides resulting from electroplating processes is the chlorine method.

An interesting method to resolve the problem of cyanides consists of a concentration procedure followed by the re-use of the concentrate due to a re-cycling process. In such a scheme, inverse osmosis would be the best method and could be used once membranes resistant to pH conditions of treatment have been found. Studies on this type of process are being undertaken, at the present time, by one of the members of the Consultative Commission responsible for the present report.

h.4 Active carbon method

Lastly, a method is described which consists of a catalytic oxidation of cyanides by the passage of effluents over active carbon impregnated with a copper catalyst or other oxidizing agent.

This method should be capable of producing an effluent containing between 0.05 mg/l, starting from an initial concentration of 20 mg/l cyanide. A primary treatment process, is however needed, in order to eliminate suspended matter and certain dissolved metals in the effluent.

i. Some general economic data on the treatment of cyanide containing effluents

i.1 For treatment installation using the **hypochlorite oxidation** method, (28) and (29), investment and running costs are given below in table 4.3.i below.

Table 4.3.i

Treatment capacity of installations (m^3/h)	1	100
Investment costs		
Total in thousands of u.c.	10.51	281
Per m^3 effluent treated of u.c.	0.70	0.14
Running costs in u.c. per m^3 effluent treated	0.92	0.32
Price cost of treatment per m^3 of effluent	1.62	0.46

i.2 Concerning treatment installations using **electrolytic oxidation**, it is possible to indicate that for an effluent containing 4 g/l of CN the electrical energy consumption is 10 kWh per kg of CN^-, this gives a total cost of about 0.27 u.c. per kg of CN^- treated.

i.3 Finally, due to the fact that **catalytic oxidation** treatment is a very exothermic reaction, the energy required for the process is very low. The total cost, per kg of CN^- treated varies between 0.18 and 0.54 u.c. depending upon whether the sulphuric acid used is new or recovered (recycled).

C H A P T E R 5

MEASURING AND MONITORING SYSTEMS OF THE DIFFERENT KNOWN FORMS OF POLLUTION FOR THE POLLUTING ELEMENTS OR COMPOUNDS CONCERNED

CHAPTER 5 - MEASURING AND MONITORING SYSTEMS OF THE DIFFERENT KNOWN FORMS OF POLLUTION FOR THE POLLUTING ELEMENTS OR COMPOUNDS CONCERNED

In this chapter, much emphasis is laid upon new technologies to prevent certain toxic substances from reaching the environment. Regardless however, of what preventive technology will be developed, continuous inspection of the water quality will always be necessary. As long as the toxic substances discussed in this report are used in production processes, are transported or are used by the public as end-products, there will always exist the risk that they escape in the effluents of production centres, in domestic sewage or become liberated, in some controlled or non-controlled way, from the treatment of domestic or industrial waste.

From the viewpoint of public health as well as of marine biology, it will be necessary to watch closely the development of the trend of concentrations of toxic substances and, if sudden changes in the concentrations are observed, an alert should be issued to those governmental and industrial bodies which are responsible for environmental protection in order to enable them to take such control actions as to minimize unwanted effects.

We will give a short survey of the available methods for the analysis of water, their limits of detection, their reliability and the possibilities for automatic operation. Prior, however, to this discussion on analytical methods, we will bring forward some aspects of the setting up of an inspection system for water quality.

5.1 WATER QUALITY

a. General aspects

a.1 Introduction

The main object of this study is the protection of the sea against some toxic metals present in industrial effluents, particularly chromium, cadmium, zinc and also cyanide-containing compounds. The protection of the sea as such, however, is impossible without taking into account the rivers, streams and coasts.

Although only a small part of the ocean water is supplied by run-off from the continents (69) (about 38.10^3 km^3/year against a precipitation of 320.10^3 km^3/year)* and the part supplied by rivers polluted by industrial activity is only a fraction of that, it is clear that oceans, inland seas,

**It is not unlikely that contamination of sea-water by pollutants in rain is an even more important source than effluents of rivers. At least for pollution by hydrocarbons, there is strong evidence that the main source is of atmospheric origin.*

coastal waters and rivers are so closely connected that inspection of water quality should be conducted in the saline part as well as the fresh-water part of the total world water volume. Moreover, most of the toxic substances are discharged into fresh-water and the inspection should be carried out as close as possible to the source of pollution, not only because of the fact that the concentrations are higher, which simplifies the analysis, but also because of the short time span required between an unwanted discharge and a subsequent alert. As an example of the influence of river water on the composition of coastal water, we will discuss for a moment the pollution of the North Sea by the river Rhine.

a.2 The North Sea/Rhine scheme

The North Sea has to take up, in the first instance, all pollutants present in the river Rhine. Because of the fact that these pollutants are relatively concentrated in coastal waters, their effect will be more pronounced there than on the high sea. The dilution-volumes, however, are already so large that only the substances with low aqueous solubility come immediately to our attention (floating waste, oil, etc.), and dissolved components can be found only by chemical analysis of sea water.

The water of the North Sea consists of two parts:

- first, the "residual-current" flowing from the Channel to the North, bringing yearly 2000 km^3 of ocean water along the east coast and superimposed thereon the tidal current, which makes the movement of the water particles not linear but spiral shaped. This "residual-current" is strong enough to force the Rhine water to follow a north-bound course and to be "pressed" against the Netherlands coast.

- the second part of the North Sea water comes from the river Rhine (approximately 70 km^3/year). During every tidal period, about 0.1 km^3 of Rhine water is discharged into the North Sea.

The power necessary for mixing two such large volumes of water with different specific densities is delivered by the tidal movement. The result is that, yearly, about 400 km new coastal water is formed, consisting of 1 part of Rhine water and 4 to 5 parts of sea water.

Due to this mixing by tidal movements of polluted Rhine water and relatively clean Atlantic water, no serious calamities have taken place up to now along the west-European coasts. The situation in the Baltic and in the Mediterranean, where hardly any tidal power is available might, therefore, be more dangerous to the marine environment.

a.3 Pollution of the North Sea by the Rhine - some figures

For acceptable limits of discharge of foreign substances, no scientifically founded criteria are available at this moment. A rule of thumb for long-persistent metallic substances is the following: the extra concentration due to discharges of persistent substances in the sea should not increase the natural concentration of the ocean by more than 50%.

This rule gives, more or less, the allowable concentrations of trace metals in the river estuaries. In the following table, the natural concentrations in the North Sea (70) (72), are listed, together with the maximum allowable concentrations in the Rhine (assuming a mixing ratio of 1 to 4), and the concentrations of these metals as observed at Lobith and Nijmegen (71). The last column gives the ratio between the computed and observed concentrations or the overload of the river Rhine.

Table 5.1.a.3-I
Concentrations of some elements in the North Sea and in the river Rhine

Element	Concentrations in μg/litre			Ratio
	Seawater	Rhine*	Observed**	
Na	105×10^5	263×10^5	0.7×10^5	3×10^{-3}
K	380×10^3	950×10^3	7×10^3	7×10^{-3}
Mg	135×10^4	340×10^4	1.5×10^4	4×10^{-3}
Ca	400×10^3	100×10^4	80×10^3	8×10^{-2}
Hg	0.3	0.7	0.1 - 6	0.3 - 9
Pb	3 - 5	7.5 - 12	2 - 78	0.2 - 10
Cd	0.1	0.25	1 - 8	4 - 32
Zn	10	25	50 - 570	2 - 230
As	3	7.5	8 - 20	1 - 3
Cu	3	7.5	1 - 75	0.1 - 10
Cr	0.05	0.13	2 - 125	15 - 1000

* *Maximum admissable concentration*
** *We will come back later to the reliability of the measurements*

Especially for the elements cadmium, zinc and chromium, the load carried by the river Rhine is much larger than desirable from the viewpoint of marine biology. It goes without saying that, for other applications of fresh water, such as drinking water, other requirements have to be met, especially by the concentrations of salt, odorous compounds, phenols, etc., which will not be discussed in this report.

As outlined above, the protection of the seas starts with the protection of coastal waters; the protection of coastal waters starts with the protection of the effluents of rivers. One could go back even further and state that the protection of rivers should start with the control of discharges of industrial effluents. Of course this is true and should be done where possible. On the other hand, however, the sources of river pollution are not found only at industries handling toxic substances such as trace metals. By far the largest amount of these metals finds its way to end-users spread out over the country and the waste products are found back in domestic and industrial waste and scrap. If these trace metals appear in a soluble form (e.g. fungicides, paints, etc.) the risk exists that they will come into the rivers by sewage, ground water and other effluents. Calamities and leaks during transport may occur.

Although very important, it is, therefore, insufficient to control only the known possible sources of river and sea pollution. An inspection system covering all surface water is necessary. Such a system should be made up of reliable instrumentation operating fast and continuously, in order to trace the presence of toxic metals in very low concentrations so as to enable timely and correct measures to prevent the spread of the pollution over connecting waters and to prevent further contamination of the surface waters. The intake of surface water for drinking water-preparation has to be postponed and farmers could be warned not to use the water for agricultural (irrigation) purposes. In the next chapters, we will go into some detail on the parameters of interest for water quality and the basic philosophy of such protection systems and come to a comparison of the technical instrumental methods available for this purpose.

b. Toxicology of water

The most important relevant parameter for water quality, of course, is the potential toxicity of the water. However, it is very difficult to define such a parameter for general use.

Toxic concentrations of different substances are different for different biological species. Due to accumulation in the food-chain, metabolic effects, etc., the criteria for toxicity are subject to strong variations. Furthermore, synergetic effects may play an important role in a positive or a negative sense. For this reason, it seems appropriate at the moment to derive toxicity parameters from the chemical composition of water. As the experience in this field increases, more reliable criteria can be defined by adding the concentrations of individual compounds with suitable weight factors. A monitoring system for water quality should, therefore, be based on chemical analysis of the water. A number of techniques are known for this and can be readily developed and applied.

Biological indicators, however, are still in the laboratory research field as far as continuous and automatic measurements are concerned.

5.2 PARAMETERS OF INTEREST TO WATER QUALITY

The quality of surface water is determined by a very large number of parameters, some of which are closely connected, and these form the input to a standard mathematical model of the stream. Parameters to be discussed in this chapter are split up into the following groups: physical parameters, standard chemical parameters, oxygen balance, eutrophication, odorous compounds, mineral oils and phenols, trace metals and organic micro-pollutants.

a. Physical parameters

Among the physical parameters are: level, flow and temperature, being the normal hydrologic measurements. But, also such parameters as solar radiation and precipitation in the area from which the river receives its water are of interest. Solar radiation is an important factor in photosynthesis. Together with dissolved oxygen and COD, solar radiation is an input parameter to the oxygen balance equations of natural waters. Precipitation data are of interest in forecasting level and flow but, as well, concentrations of phosphate and nitrate may change in the near future due to leaching of agricultural terrains. If the effect of rainfall can be calculated, an increase of these concentrations will not appear unexpectedly. Also, the turbidity of the stream may change due to extra discharge of clay and sand particles and, due to the increased flow-velocity, sludge may be upstirred from the sediment. In the sedimental sludge, large amounts of adsorbed metals and pesticides may be present.

b. Standard chemical parameters

Standard chemical parameters are pH, total suspended solids (turbidity), total ion concentration (conductivity) and some specific ions, which are present in higher-than-trace concentrations. Which ions should be measured depends on the expectation as to which ions can be present. Among these are Cl^-, CN^-, F^- and others.

c. Oxygen balance

The oxygen balance is determined by the dissolved oxygen and the BOD. As the dissolved oxygen depends on the temperature, it should be expressed as a percentage of maximum concentration. Therefore, the temperature (and as indicated above, solar radiation as well) should be measured as physical parameters. As the measurement of BOD takes too long and cannot be performed well by automatic equipment, its measurement can be replaced by the measurement of COD. This has the advantage that difficult or non-degradable substances also are found.

d. Eutrophication parameters

Among the parameters which are important to know, in view of the danger of eutrophication, are phosphates and nitrates. Eutrophication parameters are interesting only if eutrophication can occur; in other words, if the river discharges its water into a lake, a drinking water reservoir or another basin in which the current is low.

e. Odorous compounds

Odorous compounds, or compounds which may be converted easily into odorous compounds (amines) are important on those sites where the water is used for recreation, intake for drinking water, etc.

f. Mineral oils and phenols

A very rapid measurement of mineral oils is necessary in order to trace discharges of oil and to contain this oil before it has spread over coasts, borders and beaches. Measurement of floating oil, as well as emulgated or dissolved oil, is necessary. By subsequent analysis of composition, sulphur content and trace metals (for instance vanadium) the origin of the oil spill can be traced. Phenols are of prime interest in drinking water. Due to chlorination of raw water, phenols are converted into chloro-phenols, well known for their bad taste, which can spoil completely the quality of drinking water even in very low concentrations (down to a few micrograms per litre).

g. Trace metals and organic micro-pollutants

The black list adopted by the International Rhine Committee contains the following inorganic and organic micro-pollutants: organo-chlorine compounds, organo-silicon compounds, phosphor containing pesticides, organo-tin compounds and the metals mercury and cadmium. These compounds are no longer allowed in detectable quantities in natural waters. The grey list contains, among others such as oils, phenols and odorous compounds, the metals Zn, Cu, Ni, Cr, Pb, Se, As, Sb, Mo, Ti, Sn, Ba, Be, U and also CN, F, NO_3.

The discharge in surface water of components on the grey list must be avoided as far as possible. The components appearing on the black list will have first priority to be measured in an alarm system. Hence, for these components, fast and automatic instrumentation must become available.

The concentrations of components on the grey list have to be followed at a lower frequency and are to be measured in a trend measuring system. In the near future, we can expect a number of components from the grey list to be transferred to the black list.

5.3 MONITORING SYSTEMS

a. General aspects

A monitoring system, as the word indicates (monitorius = warning), has the task of warning. This warning can be a long-term one, indicating that the situation is slowly deteriorating, or a short-term alert. On the basis of the information supplied by the system, different actions can be started, in which tracing of the polluter is only one example. The two main objectives of a monitoring system are: indication of trends or slow variations of water quality and indication of sudden changes in the normal situation, which are unexpected variations exceeding the normal fluctuations of the concentration of the components under study.

b. Characteristics of monitoring systems

For these different tasks, different specifications have to be established for the parameters of operation of the system such as frequency of measurement, time delay between sampling and result of the analysis, accuracy, lower limit of detection, specificity, etc. On the other hand, there are also some common properties such as unattended operation, automatic data handling, telemetered and automatic calibration.

b.1 Frequency

For measurement of the general trend of pollution, the sampling frequency can be rather low, between once a week and once a day. Care must be taken that the sampling frequency is out of step with periodic fluctuations. This would otherwise lead to false conclusions, especially if daily or weekly intervals between sampling are chosen, because of the daily and weekly cycle of human and industrial activity. By random sampling, these natural fluctuations can be detected.

For alert measurements, a much higher sampling frequency is needed. This frequency depends, of course, on the statistical fluctuations to be expected, the density of discharge points along the river, the velocity of the stream and the density of intake points of surface water. For general parameters such as turbidity, pH, oxygen, etc., a read-out of the instruments every 5 to 10 minutes is obtainable with the existing techniques. For oil, an alert frequency of once per 15 minutes is a minimum. For heavy metals, little is known yet about the fluctuations in concentration. A sampling frequency of once an hour seems to be appropriate at the moment. For tracing of polluters, a very high sampling frequency is necessary.

b.2 Time lag

The time lag between sampling and presentation of the result of the measurement also determines the alacrity of a monitoring system. For trend measurement, the time lag is of no importance, provided the sample does not deteriorate during storage and transport.

For alert measurements and tracing of polluters, the delay in time has to be as short as possible. It is not very useful, however, to have this time much shorter than the time between successive samplings, which brings the delay time, depending on the component to be monitored, between a few minutes and one hour.

For heavy metals, this order of magnitude can be reached at the present time with several methods.

b.3 Accuracy and precision

The results obtained by single measurements of heavy metals in natural water are subject to a number of errors. The accuracy, and even the precision, certainly has to be improved.

A comparison carried out recently (73) may be mentioned here. Two samples of seawater were sent to a number of laboratories for analysis of their zinc content. Some laboratories used the same method, others used different methods. After leaving out the most extreme values, concentrations for one sample were reported of 4.9 μg/litre with a standard deviation of 4.6 μg/litre. For the other sample, the concentration was found to be 5.6 μg/litre, with a standard deviation of 5.3 μg/litre. If there would have been a difference in the zinc concentration at the two sampling points (Pacific and Caribbean deep water), it is not reflected in the reported results. Equally disastrous results were reported with simple and cheap instrumentation and with very complicated equipment. The scatter in the results is so large that differences in the method used cannot be detected.

This example shows that much improvement must be made in the methods for environmental chemical analysis. It stresses the need for automatic equipment, which takes samples every hour, or even more frequently, and presents the results a short time later. By closely following the development of the concentration of elements at the same place, measured by the same method, statistical treatment of the data and a much more precise value of the concentrations can be obtained than can be derived from a single sample.

For the standard physical and chemical parameters, the precision obtained now is in the order of 10%. The accuracy of the analysis can be improved considerably by frequent calibration of the sensors. Other errors incorporated in the result may arise from the sampling system.

For alert measurements, the accuracy does not need to be high, so long as a relevant change can be detected (with a high level of confidence) exceeding the normal fluctuations of the river and the statistical fluctuations of the analytical method.

The accuracy and precision for trend measurement has to be considerably higher, but as trend measurements are extended over longer time periods, they incorporate more samples and the precision can be high enough.

b.4 Limit of detection

Low limits of detection only are of importance for the determination of substances being present in very low concentrations, such as heavy metals and organohalogeneous compounds. For the standard chemical parameters, accuracy is of more importance and for some other compounds, as for instance, phosphates or oils, the lower limit of detection required is given by the maximum allowable concentrations of these components in surface water. For heavy metals, required lower limits of detection range from 0.1 µg/litre or lower for mercury, to a few µg/litre for zinc, which are possible to acheive with present day instrumentation.

For trend measurement, and alert measurement as well, the required limits are different. As trend measurements are also to be carried out in cases of low pollution, the limits of detection to be chosen are very low, for instance, 10% of the existing average pollution level. For alert measurement, the levels can be higher in some cases. Here, a threshold level should be detected with 95% confidence.

b.5 Specificity

A method is called specific if the accuracy of the result of the measurement of one component is not seriously influenced by the presence of other components. It is clear that instrumentation to be used in an automatic monitoring system (cf. chapter 3.3) should possess a high inherent specificity. As alert measurements will be carried out by such automatic instruments, the specificity of these instruments must be high. From case to case, specifications have to be set depending on the component under study and the levels of possible interfering components.

For trend measurements, the specificity can always be improved by a suitable pre-treatment of the sample, during which interfering compounds are removed.

In point 5.4 we will discuss the available analytical methods for the measurement of some components in water and compare them with the requirements mentioned above.

c. Manual and automatic systems

The aim of a monitoring system is to collect data on water quality in its broadest sense. In different places and at different times, water samples are taken, analyzed and the result of the measurement is used as the basis for a decision whether any action has to be taken. The system should therefore be set up in such a way that a maximum of information is collected at a minimum effort and cost. The design of a monitoring system depends largely on the questions to be solved, the pollutants under consideration, the topography of the area, the time resolution wanted, the desired accuracy, the expected range of concentration and, this is a strong point, on the availability of suitable analytical techniques giving reliable, specific data, directly comparable with data collected by other methods.

In principle, two different set-ups are in use for monitoring and control of water pollution:

- collection of water samples at different places and manual (or mechanized) analysis in a central laboratory;
- automatic on-site measurement of specified contaminants. In its most advanced form, the data are transmitted by an on-line telemetering system to a central point.

A mixed system can be considered, the so-called mobile monitoring system, consisting of trucks or boats equipped with measuring equipment. The measurement is carried out on-site but, seen as a network, the sampling frequency per station is low. Mobile monitoring can be used as a first step in setting up an automatic network and as an additional source of information on the data obtained by such networks.

c.1 Manual systems

This method is, of course, the oldest and is still in use for many purposes and in many countries. The water samples are collected in glass or plastic bottles, some primary reagents may be added to ensure that the sample is well preserved (setting the pH, filtration) and the bottles are transported to a central laboratory for analysis of trace compounds by well known standard analytical methods.

1) **The advantages** of the system are:

- although the basic equipment requires a large capital investment (a well equipped chemical laboratory), the sampling stations are cheap;

- all analytical techniques are known;

- to the number of components, the system is very flexible. If components have to be added to the system, it can be done without development of new instruments. Also the number of monitoring sites can be adapted without difficulty to the current needs.

2) To the advantages, a number of **disadvantages** appear, which make the system ill suited for active monitoring:

- the sampling frequency is very low. If water samples have to be taken a sampling frequency of once a day at every station seems already high. This can be improved by the use of mechanized samplers, which take a water sample every hour and store it under well defined conditions. These samplers have been developed and could also be used in automatic stations. We will come to this later;

- the time delay between sampling and result of the analysis is long;

- in general, the system will not give a round the clock operation;

- due to the low sampling frequency, the correlation between different pollutants and different places is lost. This means the tracing of polluters is difficult;

- for dense or large networks, many personnel are necessary to handle the enormous amount of samples. This will make the running costs very high.

Manual systems, as discussed above, will hardly be capable of serving the alert function. For the measurement of slow variations (trends), these manual methods can be used. Also, as a forerunner of automatic systems or to study the extension of automatic systems to more components or more stations, a manual system of measurement could be of much use.

c.2 Automatic systems

In automatic systems, the analysis of the water is carried out on-site. In these systems, it is not the sample which is transported but, rather, the result of the measurement which is transmitted to the central authority. As no transportation of mass is involved, this can be carried out easily and quickly via telephone, telegraph or telex networks.

The measuring instruments can be connected directly to the transmission network and data are collected without human intervention. A central data processor (a small computer), collects automatically the data from the different monitoring stations and performs calculations before presenting the data to the operating personnel. Deviations from the normal situation are detected immediately and instructions can be given to the built-in sample storage device to preserve samples for subsequent manual analysis, eventually to be used in legal evidence. The number of samples to be analyzed manually is thus restricted to those samples which have significance, which reduces the amount of work considerably. The calculations carried out by the computer may include: average values over a specified time, including statistical spread, comparison of data from different stations, checking of values against pre-set threshold levels and comparison of short term trends to expectation values so that warnings can be issued. In air pollution measurements, this system has already been in use for several years (for instance, in the Rotterdam area), with very good operational results.

1) For these telemetered automatic systems, the following **advantages** may be summarized:

- The running costs are low, especially if unmanned stations are used, connected to a well programmed data processor; one person can operate a very large network. The national air pollution monitoring network in the Netherlands may be mentioned as an example, consisting of 250 monitoring stations covering 30,000 km^2 of land area;

- time correlation between the different components and the different stations is easily found;

- tracing of polluters is quite possible;

- the system provides a momentary overview over large areas, thus enabling the computation of the best set of measures to be taken if a pollution episode is detected.

2) **The disadvantages** of the system are:

- especially for unmanned stations, the requirements to be met by the analytical equipment as to reliability and specificity are much more severe than for manned stations or for the manual sampling system. This means that analytical instrumentation has to be developed which can be used in such systems and this could make the capital investment per station rather high;

- the system is not flexible with regard to the addition of other components or other monitoring sites. Before the network is set up, a choice has to be made from the list of components to be measured, automatic equipment has to be developed, the sites of the stations have to be selected, electricity and telecommunications cables have to be laid, etc.

As has been stated above, the best system will be of the mixed type, automatic where possible for the components of prime interest and mechanized for the other components. With computer controlled automatic samplers in each station, the advantages of both systems can be combined with a minimum of cost and effort. The system will be flexible, fast and provide the maximum of information wanted for the protection of the surface waters.

d. Stations for automatic systems

d.1 General characteristics

In an automatic system for the monitoring of water quality, not all monitoring stations of the network will be of the same type. The components to be measured at a given site depend on the local situation. The measurements are not to be carried out to detect the presence of a certain pollutant only. The object is, however, how to perform the measurement of water quality in order to take the right action to prevent the pollutants from being present. This means that a certain set of measurements must be carried out as close as possible to the known pollution sources.

In general, permission will have been given to discharge a certain amount of polluting components into the river or sea. Measurement at the source is necessary to check that no more than the allowed amount is discharged and, also, to see whether other compounds, for which no permission has been given, are released.

In general, these "source monitoring stations" are of a rather simple type. Only a few parameters have to be measured, i.e. the parameters from which one knows beforehand that they are of interest here.

On the other side, there are a number of unlocated discharges. Calamities during transport, leaching from agricultural areas and industrial processes, metabolic components, etc. may affect the water quality and should be under control.

In a complete monitoring system, the hydrographic area can be divided into more or less isolated units with a known input and a known output. For these units, a computer model can be developed in which the water quality parameters are also included. The number of parameters here is determined by the expected discharges but also by the potential use that can be made of the water (irrigation, recreation, drinking water, industrial processes and cooling water), and the network is derived from the mathematical model. The measured parameters are on-line used as an input to the mathematical model to compare the data with the expected values.

d.2 Types of monitoring sites

Three different types of monitoring sites can be distinguished:

1) Nodal points

All sites in the network where one or more parameters have to be measured with a certain frequency. (The above mentioned source stations are of this type).

2) Main points

All sites in the network where all relevant parameters are to be measured for a minimum insight in the general quality of the water.

3) Reference points

A few points in the network, equipped with more general equipment, to check whether the main points still present an adequate picture of the water quality. For instance, at several main points in the network, the concentration of some heavy metals is measured. The choice made for these metals is based upon the toxicity, the knowledge about occurrence, possible sources of discharge, etc.

In the reference points, all elements in the water will be measured regularly to observe that no other elements show suddenly high concentrations so that they have to be incorporated in the network instead of some other elements, possibly, whose occurrence has decreased.

d.3 Standard network

The standard network will, thus, consist of:

1) A computer model of the hydrographic area;

2) A number of automatic nodal point stations in the neighbourhood of sewers, discharging points, high concentrations of traffic (harbours), but also at those points where water is taken from the stream to be used for some purpose.

Furthermore, nodal point stations are to be placed at the boundaries of the sub-areas in the model to provide the data on input and output of the system.

3) A number of larger stations, main points and reference points, in which a number of additional parameters are measured.

In reference points, in principle, every parameter is measured, but at a low frequency. These measurements can be performed by manual standard chemical techniques. At main points, a selected number of the most important parameters, to be determined from reference point measurements, is monitored with a higher frequency, preferably with automatic equipment.

In the reference point stations, a very large number of analyses have to be carried out. This can be done in an efficient manner only if a high degree of automation in laboratory measurement and data handling can be reached. From the reference point measurements, correlations can be observed between the different components (for instance, if many different components arise from the same source) and the relevant components can be found to be measured either at main points or even at nodal points in the system.

5.4 ANALYTICAL TECHNIQUES

In this chapter, we will discuss the possibilities for fully automatic equipment to be used in monitoring stations. This means that not every known analytical technique can be applied. Although in principle there are no differences between analytical methods to be used in the manual laboratory analysis, and those for automatic monitoring, the technical realisations may be different. The reason for this difference is the way the analytical problem is approached. In the analytical laboratory, the techniques employed are as versatile as possible. Standard analytical instrumentation is used for a large variety of determinations. Among them are: spectro-photometric methods, gas-liquid chromatography and other chromatographic techniques, X-ray fluorescence, mass-spectroscopy, emission spectroscopy, neutron activation, polarographic methods such as anodic stripping, etc.

In laboratory, the sample is modified until it is converted into such a form that the final analysis can be performed by the available techniques. This sample treatment may consist of destruction of chemical compounds containing the element under study, extraction, clean-up, preconcentration (if the concentrations are below the detection limit of the instrument), addition of reagents, etc. All these pretreatments are used to remove an interfering background and to enhance the concentration of the elements or compounds searched for. In automatic analysis to be used in unmanned monitoring stations it is not the sample which has to be adapted until it fits in a standard analytical technique; a technique must be chosen which can be modified to accept the sample directly in the form in which it enters the station. This means that those techniques which are inherently specific to the components one wishes to determine and which can operate under the very severe conditions of high backgrounds of fluctuating composition have to be developed. Most of the laboratory instrumentation is designed to operate under well defined clean laboratory conditions, where interfering components have been removed. Most of the instruments marketed today for the monitoring of surface water are slightly modified laboratory instruments and they give unreliable results. Simple instruments such as ion-selective electrodes fail to operate after a short time, because of algae and slime layers formed on the surface. Methods have to be chosen in which the addition of reagents can be avoided if possible, unless the reagent can be produced on the spot (for instance by electrolysis).

a. Methods for water analysis

In this paragraph, a few analytical methods will be described for the automatic analysis of water. In a short review, we will go over the principal methods for measurement of standard chemical parameters, oxygen balance, eutrophication parameters, oils, and, because of the subject of this Eurocean study, in some more detail over the methods for determination of heavy metals.

a.1 Standard chemical parameters

PH and some specific ions are today determined by ion-specific electrodes. From these, pH electrodes give acceptable results as to specificity and response time. For chloride, ion-specific electrodes are slow (10 minutes response time at least) and subject to interferences of different kinds. In the future, these electrodes might be improved; other techniques (coulometry) show, however, more promising properties. For the coming years, however, the Cl^- electrode is the best compromise.

Total ion-concentration is measured as conductivity. With a 4 point conductivity cell, no problems are found with respect to contamination. For other ions (CN^-, F^-), ion-specific electrodes are in use, but as the levels are in general low, their reliability is poor.

The measurement of turbidity by light scattering does not need any improvement.

According to data published by some manufacturers of these instruments, the accuracy, drift and response time of the different methods can be summarized in the following table.

Table 5.4.a.1-I

	pH	pCl	Conduct	Turbidity
Range	5-10	2-5	0-2000μS/cm	0-100 TU
Drift 24 hrs	0.05	0.05	0.1%	-
Spandrift 24 hrs	0.05	<0.1	-	-
Linearity error	< 1%	5%	1%	5%
Repeatability	<0.01	0.01	<0.1%	<1%
Response time	2 min.	10 min.	2 min.	<2 min.

a.2 Oxygen balance

The oxygen balance is determined by the concentration of dissolved oxygen and the BOD. As the determination of BOD takes at least 24 hours, it has been replaced in most cases by a measurement of COD.

Dissolved oxygen can be determined in a reliable way by a membrane amperometric sensor in the range 0 to 100% of the saturation value. Drift and span non-linearity do not exceed 1% and the response time is less than 2 minutes.

COD is determined in the laboratory by the $K_2Cr_2O_7$ method. The water sample is boiled for a few hours with $K_2Cr_2O_7$ as oxidant and the consumption of the oxidant is determined either by titration or photometry. The method is slow (a few hours), chloride ions interfere and have to be removed, and the accuracy for surface water (COD < 30 mg/litre) is not very high. Fast methods determining the TOD (total oxygen demand) are based on combustion of the combustible parts of the sample in a carrier gas flow and subsequent determination of the consumed oxygen (74). These methods are fast (a few minutes), more accurate even at very low TOD values and suitable for automation. For a rapid indication of the COD value in natural waters as well as in sewage, the UV absorption is a good indication.

a.3 Eutrophication parameters

For the determination of phosphate and nitrate, only spectrophotometric methods are in use at the present time. In many cases, this will be sufficient because of the fact that only slow changes in these concentrations are to be detected.

Mechanized versions of the photometric instruments have been constructed. Typical development specifications are:

- measuring range : 0 to 5 mg/litre,
- total phosphorus time of measurement : 20 minutes to some hours (75).

a.4 Mineral oils and phenols

For oils and phenols different methods have been developed for trend measurement and for alert. For the latter, the limits of detection are higher than for the first purpose. For floating oil, reflection of infra-red radiation is used to detect oil films down to 10 micron thickness. Suspended oil and oil in solution can be detected by fluorescence (detection limit around 1 mg/litre and not very specific), by automatic extraction of the sample and subsequent measurement in the UV bands or, after extraction, pre-concentration and clean-up of the extract by measurement in the infra-red bands (3.30, 3.38, and 3.42 microns for respectively the aromatic parts, CH_3 groups and CH_2 groups). Instead of infra-red spectroscopy, thin-layer chromatography and gas-liquid chromatography are used. These methods, although very specific and sensitive, are time consuming laboratory methods (time needed for one determination about 2 hours).

a.5 Analytical methods for heavy metals

The possible methods for the determination of heavy metals will now be treated in more detail. After all, this study is devoted to heavy metals.

The first questions which have to be answered before instruments can be compared are:

- the concentration ranges in which the instrument has to operate, and,
- what part of water sample we are interested in.

To start with the second question, there is a large difference between the total amount of metals and the dissolved part; in both parts, the element in question may appear in different forms. The dissolved part of the element may appear as cation or anion, in complexated form as an ionic or not-ionic complex. The non-dissolved part can be bound physically or chemically to suspended matter, can be present as a sol or gel or some other form of suspension. The next table gives the concentration of different elements as observed in the river Rhine*.The second column gives the

**The values given here are different from those given in the table 5.1.a.3-I. The explanation is that the values at p.4 give the lowest and highest concentrations ever found. The values here present one set of measurements during a few days when total and dissolved concentrations were determined.*

total concentration found, the third after the solid particles had been removed in a centrifuge.

Element	Total (μg/litre)	After centrifuging
Cd	2,7 - 7,2	1.1 - 3.5
Pb	29 - 78	2 - 15
Cu	29 - 74	7 - 25
Zn	110 - 570	56 - 205
Ni	10 - 17	6 - 13
Co	2.5 - 8	1 - 4.5
Cr	7 - 124	1.5 - 18
As	8 - 20	4.5 - 7.5
Hg	0.6 - 2	0.2 - 0.4

This table also gives an answer to our first question: what concentration ranges can be expected.

In general, one will be interested in the total concentration of the elements in surface water, but this makes the analysis more difficult, because of the non-homogeneity of the sample. For the determination of elements in the 10 μg/litre range, several methods are available. We will divide them here into:

. direct methods (that is, without pre-concentration),

. indirect methods (pre-concentration required).

1) Direct methods

Among the direct methods in use for the determination of heavy metals in water are well known laboratory analytical techniques such as neutron activation analysis, mass spectroscopy and emission spectroscopy. These three are multi-element methods (i.e. the concentrations of many elements are determined simultaneously), but the instruments are expensive and difficult to operate unattended.

For automatic monitoring of some (not many) elements at the same time, or subsequently, atomic absorption spectroscopy with carbon-rod atomizing furnace seems to be the best method. Recently the life time of the carbon-rod furnace has been improved considerably. Although the method in principle has to be classified as a single-element analysis (for every element the cycle of vaporisation, destruction and atomization is different and, as well, different light sources have to be used), it is possible to develop the method to an automatic operating instrument for a few (3 to 5) elements. The following limits of detection have been reported:

Element	Limit of detection (µg/litre)
Hg*	0.1
Cd	0.01
Pb	1
As	0.1
Zn	0.01
Cr	1

Moreover, the analysis is fast: within few minutes after injection the result is known.

The atomic absorption technique can also be carried out in seawater, albeit with a pre-treatment of the sample to remove the high salt concentration, which otherwise interferes with measurement.

For very clean water (drinking water), some elements can be determined by polarographic methods such as anodic stripping. First of all, however, the organic material in the water has to be destroyed, the pH has to be optimized for every element and possibly interfering elements have to be removed. Although the method is simple, its use will be restricted to the laboratory. Only a few elements can be determined. Pre-treatment of the sample by chromatographic methods (ion exchange columns) could improve the selectivity, but should be investigated further.

The last method we will mention here is the ion-selective electrodes method. The limit of detection is in general too high and the selectivity is poor. New developments (coated wire electrodes) could become more selective. For ion-selective electrodes, the same problem will occur as with the polarographic methods: choice of right environment and destruction of organic material. Contamination problems in natural water (algae, slime, sludge) are great. For drinking water, and perhaps for industrial effluents in which concentrations are high and where interfering components are absent, the method might be tried out.

**A.A.S. technique*

2) Analysis after pre-concentration

From the analytical methods discussed in the previous chapter, only atomic absorption spectroscopy could be developed to automatic instrumentation to be used in monitoring stations for surface water. A serious drawback, however, of the method is that, in principle, it is a single-element method. Multi-element methods, if the detection limit was low enough, were so complicated that their use had to be restricted to the well equipped analytical laboratory.

If pre-concentrating techniques are applied, some other methods show up. Pre-concentration can be performed on ion-exchange columns and elution with acid, by extraction and by co-precipitation. From these methods, the extraction has been developed to a point where one may expect that, in the near future, an automatic station can be equipped with instruments for a fast multi-element analysis. After extraction and concentration of the extract, multi-element analysis by energy dispersive X-ray fluorescence seems most adequate. In the example quoted above, nothing is said about the calibration techniques used. For automatic stations a reliable calibration method, which can be carried out by a remote signal, is necessary. The best calibration method would be the standard addition technique, to rule out the effects of the back ground.

b. General comments concerning seawater in particular

Going back to the report of the experts who met in December 1970 in Rome for the FAO meeting to discuss on seawater pollution, we will enumerate the following points:

. The natural concentration of the elements Cr, Cd, Zn in sea water is approximate to:

- 0.04 µg/l for Cr
- 0.02 to 0.1 µg/l for Cd
- 2.00 to 10 00 µg/l for Zn

. These micro-quantities are usually found under the following forms (more or less complex):

- hydrate $Cr(OH)_3$ for Cr*
- $ZnCl^-$ for zinc
- $CdCl^{3-}$ or $CdCl^2$ for cadmium

. These metallic elements can be dissolved or adsorbed upon the organisms or particles in suspension in the marine environment.

**The chromates are reduced in sea water:*

$$2CrO_4^{--} + 4H^+ + H_2O \rightarrow 2Cr(OH)_3 + \frac{3}{2}O_2$$

. The difficulty is that for a precise determination of the elements a good sampling and a preservation of that sample is necessary. Two things to avoid are possible adsorption on the sides of the container, or sedimentation, which diminish the value, and contamination, which increases the value.

. Common studies made in several laboratories have shown that reproduction of analyses of heavy metal traces in the marine environment are difficult (particularly non-homogeneity due to elements of marine bios).

. When sediments or marine organisms are involved, the dissolution, by continuous boiling in an acidic solution, can cause some difficulties. One must therefore be very careful of contamination by the mineralization catalysts. Current studies for mineralization at low temperature (by micro-waves) will probably facilitate this problem.

. The process leading to the concentration of trace elements is amongst others the complexion and the ion-exchange resin, the extraction by solvents and the enrichment by elimination of water. These methods are often combined. Take for example the case of extraction by solvants after concentration by evaporation.

In general only the elements which have remained dissolved under form of free-iron or of unstable complex are extracted (there is also the difficulty which can arise from precipitations phenomena during the concentration or the adsorption on the walls). It is therefore more interesting to use the atomic absorption spectroscopy (A.A.S.).

The methods of multi-elements analysis are interesting for investigation of sea water* for they give simultaneous information corresponding to an unnatural distribution of the elements, and therefore of pollution. For example, the method of extraction by dithiocarbamate pyrrolidine ammonium followed by A.A.S. or by fluorescence X multi-element, amongst others for Cr, Mn, Fe, Zn, Cd, Ph, etc.

c. Method of analysis for chrome

The ideal method is A.A.S. Generally the total chrome content is determined first, and then the hexavalent chrome.

c.1 Total chrome content

The A.A.S. is generally recognized as a method of reference. Another method accepted by the World Health Organization (W.H.O.), although less precise, is the method of photometry, after transformation of Cr^{III} into Cr^{VI} has taken place.

**One of the characteristics of sea water is the invariability of the relative concentrations of the elements.*

Oxydization of Cr^{III} into Cr^{VI} is generally effected by an excess of potassium permanganate and a subsequent reduction of $KMnO_4$ by sodium nitride; colorimetry is effected with diphenylcarbazide. This method can be mechanized, but not automated for use without monitoring. However, it must be specified that all the photometric methods are influenced by strong interference due to other constituents.

c.2 Hexavalent chrome

Cr^{VI} is extracted by a solvent and then measured by A.A.S. In the same way, and with the same specifications as mentioned above, photometry with diphenylcarbazide can be used.

d. Methods of analysis for zinc and cadmium

The ideal method here is again A.A.S.*. The solvent is a mixture of dithiocarbamate pyrrolidine ammonium and methyl isobutyl ketone. The usual atomic absorption method with flame, can be used, but one can obtain better results with a furnace without flame.

Therefore the A.A.S. method is recommended; it is taken as the method of reference for the "W.H.O. manual" on water analysis; it is also recommended by the experts of the ISO TC 147 group who are concerned with the normalization of water analyses.

There are two other methods equally used:

- the method with diphenylthiocarbazone, which is a colorimetric method;
- the method by anodic re-dissolution, followed by chronoamperometry, which is a polarographic method.

Photometry with diphenylthiocarbazone is a non-specific method (there is always an over-lapping of the extraction and absorption levels for similar elements: Pb/Bi for example, or Zn/Cd, etc.). However, to carry out analyses in a known environment (drinking water) the photometric method can be used.

For polarography by anodic re-dissolution, which is better for cadmium and zinc, this method can be equally efficient, but involves difficulties of pre-conditioning of the sample (pH).

We will briefly discuss the method by "neutronic activation"**.

**If the method of a coal furnace is used, a pre-concentration of the sampling is not necessary. In fact for zinc the concentrations are usually so high, that a direct method is always possible.*

***In a nuclear reactor*

It is a particular method, used by only a few laboratories, the duration of the analysis being long, of several weeks.

e. Methods of analysis for cyanide and cyano-compounds

After a preliminary process of separation by distillation, after acidification for the total content of cyanide, or separation by microdiffusion*, colorimetry with pyrazolone pyridine can be recommended.

Nota (1) if there are no traces involved, argentometry method can be used,

(2) if only the evolution is being observed, the use of a specific electrode is possible.

5.5 GENERAL ECONOMIC DATA**

a. Introduction

In this report a rough estimate will be given on investments and running costs for monitoring systems as well as for individual equipment for measurement. It cannot be more than a rough estimate, because of the fact that a number of parameters cannot be determined accurately as they depend largely on the particular situation.

b. Monitoring networks

For the standard parameters of water quality, (pH, conductivity, turbidity, redox potential, temperature, dissolved oxygen and one or two specific ions) non-automatic and automatic stations are possible.

The differences between the two stations are:

- in automatic stations the sensors are provided with an ultrasonic cleaner;
- automatic stations contain bottles with calibration fluids, which can be led into the measuring chambers for two-point calibration of the sensors. This calibration can be carried out by a remote signal from the central control room;
- non-automatic stations have to be visited for cleaning, calibration and data collection every two days. In automatic stations an unattended operation period of 30 days can be achieved.

**Conway method*

***Private communication*

b.1 Investments

Different firms give different prices for non-automatic stations, but it is not always clear what is included and what extra options are. A price of about 15,000 u.c. is about average.

Automatic stations with local read-out of data, but equipped with cleaning, calibration and sample storage in case of high pollution episodes will cost about 40,000 u.c. For a remote read-out about 8,000 u.c. has to be added for telemetry, coding, decoding, etc.

For the telephone extension about 100 u.c. has to be paid (in the Netherlands). A real time monitoring network, including computer and data handling will require an investment of about 60,000 u.c. for the computer and 50,000 u.c. per station. It is clear that such systems are only to be envisaged if 10 or more stations are considered.

b.2 Running costs

In the United States of America as a general rule it is taken that the cost of installation of a monitoring station (including the station), is of the order of 20,000 u.c. to 25,000 u.c. As running costs a same figure is estimated. It should be mentioned, however, that wages in the United States are somewhat higher than in Europe and that distances are, in general, larger.

The monitoring stations for surface water there are of a semi-automatic type and have to be visited every few days and the costs include the conversion of recorder strips into figures.

The running costs of a completely automatic station in Europe (visit once per month, computerized read-out of data) will be much lower.

EXAMPLE OF CALCULATION FOR U.S.A. AND EUROPE

U.S.A. *One engineer can visit 2 stations per day. Per month he has to visit every station 12 times (average period of operation 2.5 days). One month counts 22 working days, hence 44 visits or 44/12 = 3.75 stations per man. In addition to two engineers one needs one man in the workshop for repairs, etc. and one man to read the strip-charts. Thus 4 people are needed to run 7.5 stations or less than 2 stations per man can be served.*

Including travelling, spare parts, workshop, etc. one engineer will cost about 40,000 u.c. which compares very well with the estimation of 20,000 to 25,000 u.c. per station.

EUROPE *For a completely automatic and computerized network in Europe the cost estimation runs as follows:*

One engineer could perhaps visit 2.5 stations per day. Per month he has to visit every station once.

One month counts 22 working days, hence 55 visits and also 55 stations per man. In addition to two engineers one needs one man in the workshop but nobody to read the strip-charts.

Thus 3 people are needed to run 110 stations, or 37 stations per man.

Including travelling, workshop, etc. an engineer will cost about 30,000 u.c. per year, hence the running costs per automatic station will be about 900 u.c.

To this has to be added the rent of telephone lines (in the Netherlands about 3,000 u.c./year per station), which brings the running costs to about 4,000 u.c. per year per station.

The operational costs for automatic networks are thus much lower than for non-automatic networks. Although the costs are lower the information is higher. A direct real-time collection of data is achieved including a fully automatic data handling so that immediate action is possible.

To the operational costs of a monitoring network consisting of non-automatic stations, the final data-handling and storage has to be added, which can be estimated, depending on the level of sophistication, to range from 20,000 to 50,000 u.c. per year.

SUMMARY

NON-AUTOMATIC STATION IN EUROPE

Investment	*15,000 u.c.*
Write down (7 years)	*2,000 u.c.*
Average interest 10%	*750 u.c.*
Operational costs/station	*15,000 u.c.*
	17,750 u.c.

p.m. : data handling + storage 20,000 to 50,000 u.c. for the whole network, or 1,000 to 3,000 u.c. per station.

AUTOMATIC NETWORK IN EUROPE (20 stations)

Investment	*1,050.000 u.c.*	
Write down (7 years)		*150,000 u.c.*
Average interest		*53,000 u.c.*
Service costs		*18,000 u.c.*
Telephone lines		*66,000 u.c.*
Operational costs/20 stations		*287,000 u.c.*

or per station 14,000 u.c.

b.3 "Intelligent" stations

The major part of the running costs of automatic networks consists of the rent of telephone lines. In the Netherlands this is rather cheap, in other countries these costs may be two to three times higher, thus rising the operational costs per station up to 16,000 to 19,000 u.c.

The development of small process computers has made it possible to equip the stations with their own computer (dial-link system). In such a system the central processor can call once a day the monitoring station to collect the data, which are stored in the local computer memory. Furthermore the local computer can take a programmed action if some deviation from pre-set values occurs at the monitoring point (for instance call the central processor).

The extra costs of such a system are of the order of 12,000 u.c., thus increasing interest and write down by about 23,000 u.c. per year, but the saving in telephone costs can be much higher.

As far as the author is aware these types of "intelligent" stations are being fully developed but not yet installed.

c. **Measurement of heavy metals**

In principle two different systems can be used for the determination of Cr, Cd and Zn in water, namely:

. colorimetric method,

. atomic absorption spectroscopy.

c.1 **Chromium**

The colorimetric method is based on the colour formation of Cr^{VI} with diphenyl carbazide. To determine total chromium the Cr^{III} has to be oxidized to Cr^{VI} by persulphate with silver nitrate as a catalyst.

Preparations for the analysis (calibration, standardization of reagents, etc.) will take about one man day (150 u.c.); this has to be repeated every 2 weeks. An analysis time of one hour has to be accounted for (20 u.c.). If more samples are analysed in one run, the cost per analysis goes down to about 7.50 u.c. For the total chromium determination about 20 minutes (7 u.c.) has to be added to carry out the oxidation (addition of reagents, boiling, etc.).

Weak interferences of Cu, FeIII, MoVI, Hg are reported*. The method is, however, considered as rather specific.

**It should be noted, however, that although the interference with FeIII is weak, the concentration of iron in natural waters may exceed the concentration of chromium by some orders of magnitude.*

The limit of detection is about 100 µg/litre, the upper limit being about 1,000 µg/litre; above this limit a dilution step is necessary.

The whole procedure has to be carried out by a skilled analyst. The investment in instrumentation is of the order of 4,000 u.c.

The **atomic absorption method** is used to determine total chromium. The sample can be injected directly into the instrument. Including preparation, the analysis will take 15 minutes, but a run of 20 determinations can be done in half an hour.

The investment in instrumentation is of the order of 40,000 u.c., but, if a good recipe is made, the instrument can be handled by unskilled persons.

The limit of detection can be as low as 1 µg/litre.

c.2 Cadmium and Zinc

The colorimetric method is based on colour formation of Cd and Zn with dithizone (diphenyl thiocarbazone). Preparation time (cleaning of reagents, etc.) will take a few days (300 to 400 u.c.). Analysis time is certainly more than one hour.

The detection limit for Cd is about 100 µg/litre, the upper range 1,300 µg/litre. Interference by Hg is reported. For Zn the detection limit is 100 µg/litre, the upper limit 700 µg/litre. Interferences are reported from Co and Ni.

The atomic absorption method is again very direct with detection limits for Cd as well as for Zn of 0.01 µg/litres. For Zn this lower limit of detection is not very useful as the natural concentration in sea water is already of the order of 10 µg/litre.

Investments and running costs can be derived from what has been written for chromium.

In conclusion the costs of manual analysis can be estimated as follows:

Investment	4,000 u.c.	
Write down		600 u.c.
Interest		200 u.c.
One full time analyst		30,000 u.c.
		30,800 u.c.
	or	31,000 u.c.

Where one analyst can handle about 2,500 samples a year, this brings the costs to about 12 u.c. per analysis per element.

For the instrumental analysis (A.A.S.) the rough figures are:

Investment	40,000 u.c.	
Write down		5,700 u.c.
Interest		2,000 u.c.
One full time unskilled operator		25,000 u.c.
Cost of operation		32,700 u.c.
	or	33,000 u.c.

When about 10,000 samples per year can be handled, this brings the costs down to about 3 u.c. per analysis/per element.

For smaller amounts of samples per year both methods grow more expensive, especially the colorimetric method.

The effort in making standardized reagents does not diminish with the number of samples. The turnover point is around 100 samples per year with an average price of 60 u.c. per sample*. (In some cases however, there is no turnover point as the range of applicability of the two methods may be imperative).

In the cost of measurement as presented above, the costs of sampling and sample transportation are not included. For extended areas sampling could be done by local persons and transportation by surface mail; for small areas (industrial centres for instance) one or two persons can be responsible for collecting samples and bringing them back to the laboratory. In both cases the incorporation of automatic sampling and sample storage equipment is to be considered.

The costs of data handling, presentation storage and retrieval are not included, hence, the costs mentioned above are for the chemical analysis only.

**The fixed costs are arrived at by the periodic preparation of standard reagents used in the colorimetric method. This has to be done once per week, hence the fixed running costs for colorimetric analysis are as follows:*

Write down	*600 u.c.*
+ Interest	*200 u.c.*
+ 26 man days	*26 x 150 u.c.*
	4,700 u.c.

plus a cost per determination of 15 to 20 u.c.
For instrumental methods only the terms relative to interest and write down should be taken into account plus a cost per analysis of 2.5 to 3 u.c.

d. Automatic monitoring of heavy metals

d.1 Atomic absorption method

Fully automatic operating instruments for the monitoring of heavy metals seem to be in a final state of development. The actual measurement employs, in general, the atomic absorption method; pre-treatment of the sample is performed either by a mechanized version of the standard laboratory technique, or by well-known, but up till now not yet applied methods.

Concerning the cost-price of these instruments only guesses can be made. On one hand they can be cheaper than the versatile universal atomic absorption spectrometers because of the fact that they are dedicated to the determination of one element (non-programmable cycle, one atomization source, no lamp turret, no monochromator); on the other hand they may be more expensive, because of the requirements for remote operation (automatic injection, cleaning, automatic calibration). A realistic price will be somewhere between 40,000 and 50,000 u.c.

In the near future an unattended operation time of about one week can be expected for these instruments. Based on the assumptions mentioned above for automatic stations we come to the following cost estimation for an automatic network:

One engineer can visit 2.5 stations; every station has to be visited once a week. This means that one engineer can be responsible for 12 stations. On the basis of one man in the workshop for 2 field engineers, then one man is needed for 8 stations.

Service costs thus are of the order of 4.000 u.c. per year. The cost of automatic monitoring of heavy metals can thus be estimated as follows:

Investments	45,000 u.c.	
Write down (7 years)		6,500 u.c.
Interest		2,200 u.c.
Services		4,000 u.c.
Telephone lines		3,000 u.c.
Operational costs per station		15,700 u.c.

This seems to be a high figure, but one must bear in mind that this type of monitor will give at least one measurement per hour, hence around 9,000 measurements per year.

These data are immediately stored in a computer-accessible way and they can be retrieved at any moment for comparison or other computations.

Per determination the costs are about 1.8 u.c. or roughly half the price of the manual analysis by atomic absorption. Although the price is lower, the availability of data is much higher. A round-the-clock monitoring is achieved with automatic data storage. No extra costs are involved concerning sampling and sample transport (these costs were omitted in chapter 5.5.c).

Of course the combination of heavy metal monitoring with monitoring of standard parameters is obvious. A station with standard parameters and one heavy metal would thus cost:

Investment

. Station for standard parameter (including network and data handling	52,500 u.c.
. Heavy metal monitor	45,000 u.c.
	97,500 u.c.

Write down (7 years)	14,000 u.c.
Interest	4,900 u.c.
Service costs	4,500 u.c.
Telephone lines	3,000 u.c.
	26,400 u.c.

If more heavy metal monitors are installed in the same station the price will not be proportional to the number of monitors, as several functions can be combined.

d.2 Automation of colorimetric methods

A few words have to be said on the automation of colorimetric methods. Because of lack of specificity, colorimetric methods can be used only in industrial wastewater, where the possible interferences are known as the composition of the sample will not vary between wide boundaries. Therefore these methods are suitable for local stations where they can be perhaps somewhat cheaper than the advanced methods described above.

No data on costs are available. However, it seems very desirable that reliable, cheap and simple automatic modifications of colorimetric methods should be developed for use in industrial waste water streams.

e. Conclusions

As can be seen from the foregoing, manpower is always the most expensive part in monitoring and measurement systems. Increased degrees of mechanization and automatization always reduce the costs per analysis.

The extra investments (write down and interest) made to replace one man, can be of the order of 150,000 to 200,000 u.c. For this extra amount of money fully automatic instruments can be expected giving reliable, round-the-clock and computer accessible data. The use of this type of instrumentation should therefore be preferred, on an economic basis.

Of course, the prices mentioned here are only rough estimates and large deviations may occur in practice.

CHAPTER 6

THE ECO-TOXICOLOGICAL RISKS INVOLVED

CHAPTER 6 - THE ECO-TOXICOLOGICAL RISKS INVOLVED

6.1 INTRODUCTION

To qualify,
quantify,
and prevent anthrogenic degradation of the marine environment and its attendant consequences for man is one of the new challenges of our time, which is made more difficult because of a lack of understanding and knowledge of the sea.

This challenge shown in the three recommendations below, contains major divisions which should not be considered separately as each one depends on the other two; all will benefit from contributions which define more closely the problems involved.

The present report's theme is essentially one concerning preventative techniques applicable to several clearly defined pollutants. This chapter, which discusses the eco-toxilogical risks involved with each of these pollutants is not written as a specifically in-depth scientific analysis of these known techniques, but as a possible method of knowing the fate of a given pollutant in the environment and the consequences of its action on the biosphere and thus on man.

This chapter will present, among other topics:

. a schematic model of the migration of a pollutant with, at each stage, a description of the toxic impact factors (cf. figure 6.1 below),

. a more detailed discussion of the more important factors.

6.2 GENERAL CONSIDERATIONS

The point of impact of all actions on the aquatic environment is essentially situated at interfaces, that is to say, where the lithosphere, atmosphere and biosphere interact with the hydrosphere.

This point (cf. cell image) may be thought of (gross simplification) as a dilute solution containing a large variety of mineral and organic compounds, dissolved or in suspension. Situated within this medium are many flora and fauna species.

It may be separated into interacting dynamic systems for which there is no real equilibrium. Reactions, notably non-biological, tend to try and find equilibrium but constant factors cannot be maintained because of the continuous influx of material and energy.

Given the natural tendency towards stability (connected with the "invariant" factor of time) and not one of actual equilibrium, dynamic and kinetic forces are needed to explain such phenomena.

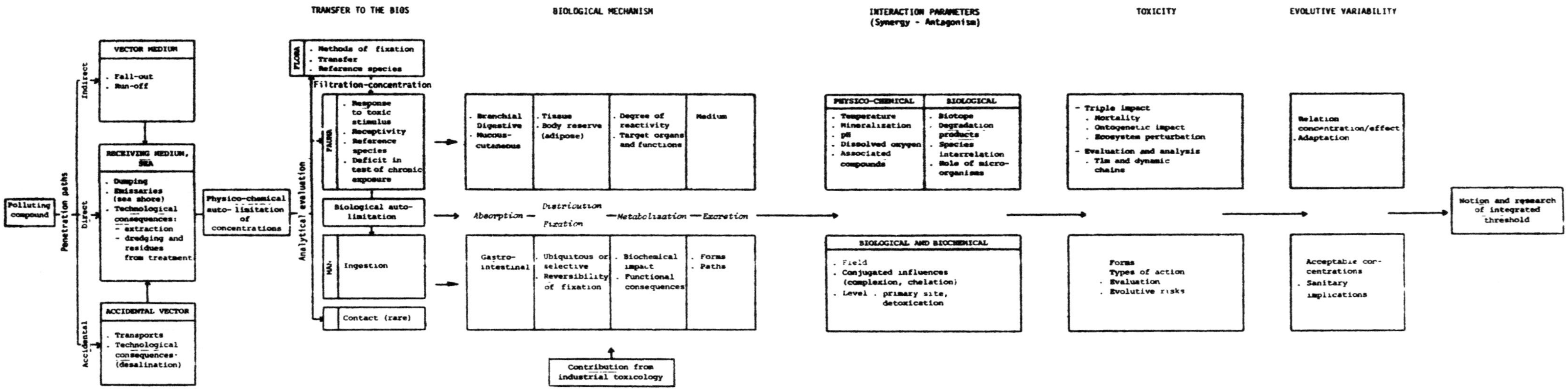

TRANSFER TO THE BIOS
BIOLOGICAL MECHANISM
INTERACTION PARAMETERS
(Synergy - Antagonism)
TOXICITY
EVOLUTIVE VARIABILITY
Polluting compound
Penetration paths
Indirect
Direct
Accidental
VECTOR MEDIUM
. Fall-out
. Run-off
RECEIVING MEDIUM, SEA
. Dumping
. Emissaries (sea shore)
. Technological consequences:
- extraction
- dredging and residues from treatment
ACCIDENTAL VECTOR
. Transports
. Technological consequences (desalination)
Physico-chemical auto- limitation of concentrations
Analytical evaluation
FLORA
. Methods of fixation
. Transfer
. Reference species
Filtration-concentration
FAUNA
. Response to toxic stimulus
. Receptivity
. Reference species
. Deficit in test of chronic exposure
Biological auto-limitation
MAN
Ingestion
Contact (rare)
. Branchial
. Digestive
. Mucous-cutaneous
. Tissue
. Body reserve (adipose)
. Degree of reactivity
. Target organs and functions
Medium
Absorption — Distribution Fixation — Metabolization — Excretion
Gastro-intestinal
. Ubiquitous or selective
. Reversibility of fixation
. Biochemical impact
. Functional consequences
. Forms
. Paths
Contribution from industrial toxicology
PHYSICO-CHEMICAL
. Temperature
. Mineralization
. pH
. Dissolved oxygen
. Associated compounds
BIOLOGICAL
. Biotope
. Degradation products
. Species interrelation
. Role of micro-organisms
BIOLOGICAL AND BIOCHEMICAL
. Field
. Conjugated influences (complexion, chelation)
. Level . primary site, detoxication
- Triple impact
. Mortality
. Ontogenetic impact
. Ecosystem perturbation
- Evaluation and analysis
. Tlm and dynamic chains
Forms
Types of action
. Evaluation
. Evolutive risks
Relation concentration/effect
. Adaptation
Acceptable con-centrations
. Sanitary implications
Notion and research of integrated threshold

Biogenic and bioactive elements (metallic compounds) are the most obvious examples of unstable distribution. The distribution cannot be explained without recourse to kinetics: this underlines one of the dominating factors: **variability.**

6.3 SUPPLY ROUTES AND SELF-REGULATION OF CONCENTRATION IN THE MARINE ENVIRONMENT

a. Supply routes

Of the various supply pathways indicated (table 6.1 above), two play an essential role:

- **transfer by atmospheric fall-out of the marine environment,**

- **land run-off from rivers, streams, pipelines, etc.**

The ecological impact may show itself in various ways (although a little over-simplified because the intricacies of the effects make analysis difficult):

. the first pathway is one of global pollution, and is applicable for areas of geographical scale. It interacts with a small biomass (the global productivity is low in ocean waters), which makes it more difficult to measure its effects (notably at phyto-planktonic level).

. the second pathway is one of regional pollution, more directly linked to industrial activity. The zones influenced are mostly continental waters and enclosed seas. The risk of toxic pollution is much greater because these zones are of high productivity and biomass.

Tables 6.3.1-I and 6.3.1-II give some information on natural and artificial materials (compounds) found in the marine environment.

Without minimising the role played by the atmosphere as a vector of transport, the most important vectors of transport of pollutants are from land run-off and direct discharge into the sea (dumping).

Table 6.3.1-II
Natural concentrations of some elements in the marine environment

Elements considered	Chromium	Cadmium	Zinc	Cyano compounds
Natural Concentration µg/l	0.04	0.02	2	-

Table 6.3.1-II
World production and potential transport to the marine environment, in thousands of tons

Elements considered	Chromium	Cadmium	Zinc	Cyano Compounds
Mineral production	2,000*	16.4	5,000	?
Pathway	run-off dumping	run-off fall-out	run-off dumping	run-off dumping
Transport by run-off	40	0.5	700	?
Part played by fall-out	20	1	-	?

b. **Self-regulating processes**

Sea water appears to finally retain only a small proportion of the material which is brought to it by rivers.

Thus, environmental processes exist within the marine environment that limit the concentration of most elements.

Of the few known (cf. Riley and Skirrow) these processes are of two main types:

b.1 Physico-chemical processes

Various reactions can take place between substances, dissolved, in suspension or in the colloidal state:

- neutralization
- ion exchange
- complex formation
- co-precipitation
- sorption of ions

This last is more likely by the dispersion of suspended matter.

**Potential chromium*

It seems, moreover, that the quantity of the absorbed element is often increased with the atomic mass in sea water; it is likely that the concentration of heavy metals is limited in this manner (the absorbing particles finish by sedimenting out of the sea water).

b.2 Biological processes

(cf. absorption - fixation, see below)

Many of the living organisms in the sea have the ability to selectively concentrate certain elements; on mineralization of organic matter, most of the elements concentrated by the organism pass back into solution, a fraction is lost by sedimentation and passes out of the cycle.

The breakdown of particles in suspension is nearly complete at a depth of 300 m (1000 feet).

6.4 TRANSFER TO THE BIOS

a. Marine flora

This consists essentially of algae, being the first step in the food chain where fixation occurs of metal compounds found in sea water.

Three points of special interest may be mentioned due to their behaviour towards this type of pollution:

. methods of fixation,

. role in the transfer to higher levels in the food chain,

. possible use as an index (indicator organism) of the degree of pollution.

a.1 Methods of fixation

The greater part of exchanges (nutritive or non-nutritive) take place, not in the substrate (tissues) but at surfaces swept by water (stipes and filaments).

A certain number of oligo-elements are required in the life cycle of a plant, of which some are present as metal complexes, either in structural components or in figured elements (thus it is found that zinc has a concentration of 100 mg/kg dry weight in a crush of ascophyllum nodosum and 500 to 750 mg/kg for laminaria).

Living in such a biochemical environment, algae are able to fix metal compounds and metaloids such as those under consideration in this report (metal complexes are just as easily absorbed as simple ions).

1) Chromium

. Standard level of fixation: 1 ppm dry weight material; concentration factor compared to water: 100 - 1,000 (for Cr^{VI}).

. There seems to be a relation between the concentration of iron and chromium (although no significant interpretation can be made).

. The average concentration of chromium in estuarine algae in industrial countries (temperate zone) is about 1.4 ppm.

2) Cadmium

. Concentration factor about 1,000 (for Fucus and Porphyra).

. Concentration for algae in exposed coastal waters: 0.25 mg/kg wet weight, this concentration may be multiplied by 100 in polluted estuarine regions.

. There appears to be a competitive relationship between the concentrations of zinc and cadmium.

3) Zinc

. Concentration factor: 10^3 (for Fucus)

. The relative accumulation tends to lead to an upper limit, this may be taken as an indirect indication of "recognition" by the plant of the metabolic significance of this element.

4) Cyano-compounds

. Data that are available at the present time are somewhat fragmentary.

. Their impact (affected by local physio-chemical parameters such as pH, temperature, dissolved oxygen, etc.) on flora is less certain than for other organisms such as fish. Inhibiting effect on growth: between 10 and 30 mg/l (Scenedesmus).

The different measurements carried out up till now have been essentially on coastal benthic flora: only a little information exists concerning phytoplankton, especially oceanic species.

Finally, a series of comparative analyses carried out over a period of 10 years, seem to indicate only a small variation in concentration-fixation of trace elements, considered in this report, by coastal sea-weeds (this does not discount possible quantitive or qualitative effects on the various species within the sea-weed population).

a.2 Transfer up the food chain

This has special significance, especially in appreciating the origins of fixation in herbivores, the results are far too scattered and few to draw any significant conclusions. It should be noted however, that certain species (benthic faunae, lamelli branches) fix more strongly from phytoplankton than from sea water (this is particularly true for zinc).

a.3 Use as an indication of the level of pollution

As with all living things in nature, the major difficulty is in comparing the tests and the results:

. each zone has its own special characteristics;

. the various species behave differently when confronted with danger;

. biological cycles and periodic changes in the structure of the plant do exist which lead to added inconveniences.

Nevertheless, the collection, the relatively easy laboratory manipulation and the importance of the fixed population allows an important understanding to be gained, based on the following tests:

. rate of growth

. modification of the biological cycle

. bio-count - qualitative (types of species)
quantitative (number of individuals)

. measuring of pigments concentration.

With regard to the above, and in the context of pollution by metallic compounds, the following species seem to be the most useful: Fucus, Porphyra, Aegregia, Scenedesmus (for cadmium).

a.4 Level of bio-activity

Table 6.4.a-IV below, shows the levels of bio-activity of the elements being considered in this report.

Table 6.4.a-IV

Element	Growth		Species measured
	Stimulating effect mg/l	Inhibiting effect mg/l	
Chromium		0.5 0.1 1.0	Scenedesmus Navicula Macrocystis
Cadmium		0.1 10 - 120	Scenedesmus Others
Zinc	0,005	0.5 - 5.0	Chlorella Scenedesmus Macrocystis
Cyano compounds		20 - 30	Scenedesmus

b. Marine Fauna

The extreme diversity of marine fauna is mirrored in the diversity of their receptivity and response to all forms of pollution (bio-agression) by metallic compounds. The different biotopes and the difficulty of making comparisons of measurements have to be underlined. The greater part of experiments may be related to the following regions:

- estuaries,

- coastal waters,

- ocean waters.

Mention being made whether these regions relate to enclosed seas or to oceans.

In these regions, the life characteristics are so different that often further distinctions have to be made between:

- fixed organisms (corals, hydroides),

- sedentary animals (molluscs),

- free moving animals (migrating fish).

The different types of food dependency add a further variable to a picture already variable. Thus, in regard to the limited understanding, at the present moment, one must read the information presented here in its context, as purely indicating an approximate outline for the picture.

Four headings may be indicated:

b.1 Response of a species to a toxic stimulus

For the majority of species (chordates), the detection ability of their senses (whether it be an intermediate neuro-sensor or chemical mediator) allows them to avoid water masses containing high concentrations of toxic substances; certain fish are capable of detecting concentrations as low as 10^{-5} mg/l. Scientists have even attempted to determine the corresponding sublethal concentrations; nevertheless, this avoidance is not without effects on the equilibrium of the biotope, for example alteration of the migratory behaviour.

In fact, the majority of toxic compounds (and more specially metallic compounds) are not detected, at least not before a certain degree of sensibilization; the influence is thus an insidious one affecting their behaviour and, even in the absence of metabolic toxic impact, may interfere seriously with the physiological balance.

b.2 Organic receptivity to metallic compounds

(cf. fixation)

The receptivity of a species is completely variable and depends as much upon the specific species, as upon stable chemical form of the toxic compound considered.

1) Chromium

. A natural element present in numerous organisms in varying concentrations:

approximately	1 ppm	(dry weight) in lamellibranches
	10 ppm	(dry weight) in worms
	<1 ppm	(dry weight) in fish

. its fixation seems to depend upon simultaneous occurrence of other metallic elements (notably iron).

. its transfer factor (or concentration factor) from water seems to vary between 1,000 and 3,000 (at least for Cr^{III} which is rapidly absorbed); and for Cr^{VI} between 40 and 100 (with a longer factor for molluscs).

. fixation seems much greater (clearly so) for molluscs than for fish (but there are also large variations between species: from 1 to 10 µg/kg wet weight).

. the lethal range varies between 20 and 200 ppm.

2) Cadmium

. Present in nature, in non-polluted sea water between 0.01 and 0.1 µg/l.

. it is present in the free or solubly complexed form (chlorides, sulphates or nitrates).

. its concentration in various species is variable, one can generalize the average concentration as being:

- coastal waters undergoing pollution:	0.5 µg/l
- polluted estuarine waters:	2.5 µg/l
- polluted estuarine sediments:	1 to 3 mg/kg (dry weight)
- molluscs from polluted estuaries:	10 mg/kg (wet weight)
- non-sedentary fish: (exposed to continental and estuarine waters)	0.25 mg/kg

. average transfer factor from water between 1,000 and 10,000,

. certain workers have put forward a tolerance limit in sea water of 40 µg/l (interpretation very hazardous).

. it plays a competitive role with zinc.

. the lethal range varies between 0.1 and 1.0 ppm.

3) Zinc

Zinc is an element whose metabolic reactions are directly influenced by the presence of other associated elements, as well as its own physicochemical form. Cadmium, especially, seems to restrict the absorption of zinc without, however, modifying the part already fixed.

. it is particularly strongly absorbed by lamellibranches (but here also there are big variations between species, from 1 to 100 times),

. its transfer factor from water varies between 500 to 50,000,

. encrustation of certain lamellibranches at 0.2 ppm.

. one finds, however, a variety of responses of stunted growth of 'tests' at concentrations of approximately:

. 1.0 mg/l for Artemia

. 0.1 mg/l for Polychaetes

. it fixes preferentially at the excretory organ (approximately 2/3 of its total fixation).

. the lethal range varies between 0.2 and 10 mg/l.

4) Cyano compounds

. not found in the natural state,

. their toxicity depends upon pH (weak at pH 8 and strong at pH 6),

. they are very often rapidly bio-degraded,

. the lethal range varies between 0.01 and 0.5 mg/l.

b.3 Indicator organisms

From the experimental data available, the results underline:

- the large receptivity of molluscs and certain benthic (worms) or fixed organisms (hydroides) for metallic compounds,
- the greater importance of filtration through gills compared with the digestive uptake,
- a concentration variability within a single species (cf. interaction effects below).

The difficulty of assessing the relative nature of the responses.

However, in the context of metallic compounds, the following experimental tests should be mentioned:

. **Environmental observation**

- distribution of species and individuals (polychaetes) (1)
- changes in habitat (flat fish),
- changes in behaviour (avoidance, swimming),

(1) Measurements which take into consideration natural fluctuations (yearly cycles, species compensation, etc.) are relative.

. **Macroscopic observation**

- rate of growth (polychaetes, hydroides, crabs),
- respiration rate and ciliary (vibration of hair-like organs) activity,
- deformation of shell and carapaces,
- disturbance of embryonic development and metamorphosis,
- osmo-regulation tests.

. **Microscopic and quantity observation**

- concentration in haemoglobin,
- degenerate cells (hepato-pancreas),
- enzyme activity.

b.4 Deficiency of tests employed for measurement of chronic exposure and long-term effects

Experimental reproductibility of stimulation and toxic exposure tests are usually carried out over a period of 96 hours, occasionally up to 3 weeks and very rarely for periods up to 50 days.

It is practically impossible to determine the real impact in such short periods, especially of long-term effects. These effects, especially on the qualitative and quantitative distributions of various biocoenoses, represent the major danger to the marine environment. Too few elements exist which are able to give information on reproductive, embryonic and larval changes, on pathological sensitization, on degenerative processes.

In the marine environment, there is no discrimination possible, and the tolerance norms must be applied, not to the adult population but to the young which are more sensitive and thus most at risk.

6.5 BIOLOGICAL MECHANISMS (ABSORPTION, DISTRIBUTION AND FIXATION)

The effect of filtration-concentration of marine species is governed by five mechanisms:

- ingestion of suspended particles,
- ingestion of pre-concentrated nutritive elements,
- chelation of metal ions by organic chelating agents,
- incorporation of metallic ions in the physiological systems
- metal ion capture on mucus membranes.

a. Gills

Dissolved substances (such as heavy metals) penetrate the gill lamelles. Two types of parameters come into play:

a.1 Physico-chemical

The gills are similar to a selective, semi-permeable membrane:

→ . degree of dissociation of dissolved substances,

. characteristics of surroundings : - temperature,
- concentration of various ions (Ca, K, Mg....).

a.2 Biological

The organic condition of the gills, at the time of contamination has an influence on the penetration, a sub-ulcerative stage or a warping of the epithelium injures the capacity to filter selectively and allows entrance of certain dissolved substances in much greater quantities.

a.3 At the metabolic level, disturbances concerned with hematopoiesis and respiratory pigments.

b. Digestive level

The role of absorption, of lesser importance in normal conditions becomes critical if the food is made up from an already contaminated alimentary chain. It is then, by the normal phenomena of concentration that the concentration factor increases, in proportion to the number of trophic levels. The metabolic impact concerns, most closely, the neuro-endocrinic system and enzyme activity.

c. Cutaneous-mucus level

c.1 Mechanical action

Irritation and warping of the gills.

c.2 Bio-chemical action

- fixation-absorption at interfaces for ions,

- fixation by mucus (especially for colloids).

c.3 Functional action

The chemical senses of fish relate for the most part to cutaneous organs: that the sensing is of a physical nature (nervus lateralis) or by a bio-chemical intermediary, the presence of dissolved toxic substances (for example cadmium) causes a change in the behaviour. This change is much more of the nature of a disruption of inter-relationship than of escape.

Certain tests (based on these various pathological influences) have been proposed:

- determination of basic metabolism (example, shrimps), this test shows little of significance,
- variation of blood components,
- body and nerve conductibility (nervus lateralis).

In practice, it is difficult to draw any definite conclusion.

6.6 INTERACTION PARAMETERS

The marine environment comprises a mosaic of sub-systems, each having its own specialization and which interact within varying geographical frameworks (local, regional, etc.). Each sub-system is, at any given moment, the result of the influence of a large number of parameters which, in the natural state, (free from external stress) automatically compensate any tendency towards unstable equilibrium.

The introduction of a supplementary factor (example metal compounds) will not only have direct effects, but equally well indirect effects that will modify the resulting toxic impact.

These parameters are of two types:

a. **Physico-chemical**

a.1 Temperature not only affects density, evaporation and solubility, but also the bio-chemical reaction rates (metabolic, enzyme reactions...); a general rule is that the absorption and toxity of a given pollutant (cyano-compounds and cadmium compounds) increase with increasing temperature, even small rises in temperature (the toxity of cyanides doubles for each 10° temperature rise).

The temperature of marine waters varies less rapidly and with less amplitude than in fresh waters. The flora and fauna of marine waters are thus less tolerant to changes in temperature. It is in coastal and estuarine waters where the danger is greatest from the effects of harmful substances.

Depending upon the geographical area considered, the species show an adaption to temperature, but this tolerance is gradual throughout the year (a species may tolerate an annual variation in temperature of 20°C, but will not be able to survive a temperature change of 8°C occurring within a week).

Estuarine populations (especially those of commercial importance) do not tolerate, in general, temperatures greater than 32°C (whatever might be the temperature to which they are acclimatized). Thus in certain inter-tropical regions, the temperature peak in summer is very close to their tolerance limit.

Temperature is one of the most important regulating processes for physiological phenomena. It affects the intensity of bio-chemical reaction, enzyme functioning, molecular mobility and membrane exchange processes. It regulates not only the metabolism, but also migration, reproduction and development. Thus synergistic reactions (eg. increase of temperature with the impact of a metallic compound) will strongly increase the average toxicity (an increase of 10°C is known to double the gill passage of certain chromium salts).

a.2 A decrease in the **rate of dissolution of oxygen**, brings about an acceleration in the respiratory rhythm and thus favours, by filtration by the gills, an increased absorption of toxic materials (cadmium).

a.3 The pH influences the toxicity of substances by controlling the degree of dissociation of toxic compounds (in general, non-dissociated forms are more dangerous than ionised forms).

A range of tolerance has been proposed between: $6.7 < pH < 8.5$.

a.4 Mineral salts. Calcium salts increase cuteneous-mucous absorption and increase the fixation of zinc.

b. Biological

b.1 Quality of biotope. The environment will more easily cope with exposure to a given concentration of toxic materials if it has conserved the free use of its functions; conversely, and which is very often the case in estuarine areas, the natural ability of auto-compensation is overwhelmed and side-effects will be amplified.

b.2 The simultaneous presence of several toxic substances will:

- magnify the toxic impact (this is generally the case for various combinations associated with Zn, Cr, Ca, and Fe).

- decrease the effect by competition and opposition (example, cadmium-zinc, complexes).

b.3 Degradation products react in two ways:

- complexation with organic material and subsequent strong metabolic breakdown,

- physico-chemical potentialization stress modifying "information" at the level of species interelationships.

b.4 Influence of diseases (myco-bacteria, virus, parasites, degeneration processes).

b.5 Role of micro-organisms

Understanding of micro-organisms (especially bacteria) is still too fragmentary to judge the importance of their role in the marine life cycle. It is difficult to know their influence in the case of pollution.

Without understanding their exact role, it is possible to make a certain number of suppositions:

- Both with sediments, as well as suspended particles, bacteria may influence the formation of complexes (eg. organo-metallic complexes) and also other structures; such is the case of mineralization of living matter, the sedimentary fraction leaving the cycle, undergoes alteration.

- For certain species, the character of bacterial symbiosis, saprophytism or parasitism modifies the metabolism. In the case of certain algae or hydroides, the concentration-fixation of a given compound can only be interpreted when consideration is given to an associated bacterial mechanism.

- The influence of bacteria is far from being negligible in many enzyme processes. The possibility, recognized in pharmacology for certain of these organisms, to elaborate neo-structure metabolites by using an external stimulus, may equally be of use in the case of chemical pollution.

To understand the role of micro-organisms thus becomes an important feature in the overall understanding of marine pollution.

6.7 TOXITY OF ELEMENTS OR COMPOUND POLLUTANTS UNDER CONSIDERATION IN THIS REPORT

a. Impact

In all the relationships of bio-chemical stress by toxic compounds, different effects have been noted; their biological consequences may be grouped under three headings:

a.1 Toxic action on the bios by:

- disruption of physiological organs or systems,

- mass mortality,

- decrease in reproductivity.

a.2 Ontogenetic action, i.e. action of first stages of life (embryonic, larval, etc.). This phenomenon is particularly important in spawning areas.

a.3 Subtle and long-term action causing stress and disruption to:

- the level of primary production and of trophic relations,
- the biological cycle, basis of equilibrium,
- the stability of the ecosystem.

Although it can affect any of the three headings mentioned above, pollution by metal compounds acts most strongly on the third level. This becomes even more important as its action is felt most strongly in continental waters which contain the greater part of the marine biomass.

b. Analysis and evaluation

In total, the measures and experimental tests try to determine the permissible concentrations which will allow the normal development of fauna and flora. This aim must not limit itself simply to a short term calculation of profit (effects on the fishing industry for example) but above all, consider the overall risk of extinction of species; not only those species of obvious commercial interest but of all species. The strong interdependence of trophic levels is sufficient proof of a need for this wider approach.

This approach is, at the present level of understanding, impossible to achieve; the margin for manoeuvering is small and is centred on the following areas:

b.1 At "in situ" level by:

- understanding the evolution of populations (notably benthic), this is in practice difficult to define;
- statistical control of concentration-dilutions in sea water; due to the extreme difficulty of obtaining specific and comparative results, in practice, no real extrapolation of the impact on the bios can be made.

b.2 At the laboratory level by:

- measurement, for a given compound and species, of the level of fixation in a given area;
- short term experiments, which, because of the variability of results (due to the difficulty of maintaining constant experimental parameters) will not permit precise comparison but may give a general indication of behaviour;

- experiments attempting to reconstruct the tropho-dynamic chain (model simulating exposure). These are useful at present for estimating the possible effect of toxic materials on the various trophic levels, although the results are incomplete.

Based on these various data, the most reliable and reproducible are those of determinations of the mean limits of tolerance (TLM) sometimes called mean lethal concentrations (often given as mg/l or ppm).

In order to measure quantity and interprete the effects of chronic pollution by metal compounds, it is possible to define certain priorities: the setting up of a multi-level food chain dynamic model, based on species representation of coastal and estuarine waters applied to investigation of sea waters exposed to industrial effluents in a defined geographical area.

c. Summary table

Table 6.7.C-I below, gives the upper concentration limits of various polluting elements as a function of three different uses of marine waters (regulation of Peruvian government).

Table 6.7.c-I (56)

Class	Uses of marine waters	Elements			
		Chromium mg/l (Cr^{VI})	Cyanides mg/l CN^-	Cadmium mg/l	Zinc mg/l
I	Bathing area, Desalination of drinking water	0.05	0.01	0.01	5
II	Conservation of fauna	"	"	"	"
III	Industrial and navigation purposes	5.0	-	-	15

Table 6.7.c-II gives the permitted levels of concentrations in sanitary waters (USSR).

Table 6.7.c-II

Element	Permitted concentration limit
Chromium CrIII CrVI	0.5 0.1
Cadmium	0.01
Zinc	1.0
Cyano-compounds (simple and complex)	0.1

(Sanitary regulations index USSR)

6.8 EVOLUTIONARY VARIABILITY - PHENOMENA OF ADAPTATION

This occurs at various levels:

a. Form in the environment

Metallic elements found in the marine environment can occur in various forms: dissolved, in suspension, colloidal...., although this classification is somewhat arbitrary as there is a continued movement from one form to another (in solution, defined as less than 1/1000 of micron).

All material in suspension, (its biological role is of the utmost importance) is fractionated and thus increases its physico-chemical reactivity.

The role of sediments, formed by the deposition of suspended matter, cannot be understood without knowing the bio-chemical processes which occur during their migration. Also essential is the role of dissolved elements (notably on plankton). Thus, to relate a toxic impact to a surface layer of seawater requires a precise understanding of all the measurement characteristics..

b. Variability: time/concentration - transport mechanisms - biotopes

b.1 Time/concentration

At a given instant in time, the toxic menace of a unit mass of polluted water on the bios depends, at the same time, on a considerable number of parameters; physico-chemical nature of the compound, degree of dilution, mixing and circulation of water masses, surface or deep water, sediment characteristics, probability of meeting, etc.

Each of these parameters, is itself a result of various fluctuating factors (the temperature-salinity diagram is the principal method for measuring the mixing and circulation of water masses).

Each parameter has to be treated as a variable whose value, at the moment of measurement, is the result of the interaction of other parameters, themselves variable in nature.

In this regard the relation: Time/Concentration depends upon a "variable spatial-time relationship having multiple coordinates", and therefore it explains the extreme difficulty encountered in the building of models (mathematical) which are of limited practical use.

b.2 Transport mechanisms

Toxic compounds may be transported either in the dissolved state, or associated with suspended matter (the distinction is slightly arbitrary as there is a continual exchange from one state to the other).

Suspended matter, as well as being of vital importance for the majority of marine organisms, offers a large interface with water, which considerably increases its reactivity.

Sediments (deposited suspended matter) are of the utmost importance to the bios, especially in the aspect of pollution. It is not possible to understand their role without considering all the physico-chemical and bio-chemical processes which occur during their slow descent to the bottom. Here again the data demonstrates large variability.

b.3 Biotopes-Biocoenoses

Evidence has demonstrated the great diversity of the marine environment and it is thus impossible to appreciate the effects of a specific pollution except in a restricted sense: each area has its own "individual profile" which has specific characteristic reactions to pollution stress.

The accumulation of experimental and monitoring data can help to establish a statistical range but does not lessen the fact that each region has its own evolving reactivity (this is especially true for continental waters), and thus that any generalized regulations will have, depending upon the region, varying consequences.

c. Adaptation

This occurs at two levels:

- auto-education of species in avoiding polluted areas,

- stimulation of cellular bio-chemical adaptive processes.

The extreme diversity of the marine bios and the lack of knowledge about the experimental physiological determination, make it difficult to assess any prospective evaluation.

From a purely theoretical approach and without taking into consideration the idea of adaptation time, the following hypothesis may be postulated:

c.1 Chromium

The limit of acceptance seems to start at the level of invertebrates; although this is speculative, a process of infra-toxic fixation with auto-elimination (self regulating) may be taking place.

At the level of reproduction and muta genetic processes, the problem remains unsolved.

c.2 Cadmium

There is no indication of progressive adaptation. The competitive role of cadmium, in the presence of other metallic elements, and its accumulative potential seem likely to rule out such a process.

c.3 Zinc

Its natural metabolic role tends to favour adaptation, rather than its interaction with other elements (metallic and mineral: Ca). If it is confirmed that it reacts principally via the formation of insoluble compounds, an enzyme adaptation route cannot be excluded, in spite of an apparent selective action on haemopoietic tissues.

c.4 Cyano-metallic compounds

Their action is acute and their neutralization rapid which tends to minimize the immunizing influence. The problems corresponding to mutagenetic processes remain unknown.

To summarize, the adaptation of species (especially fauna) to repeated chemical stress has to be selective; enzyme evolution may lead to physiological modifications (especially metabolic) which could alter their nutritional or organoleptic quality as a food source. Such a possible alteration would be purely accidental (no matter when it happened) as all neo-induced adaptations occur in order to eliminate or reduce a stress rather than to cause its fixation.

6.9 TRANSFER TO MAN

The discussion here, is only presented as an outline of the problem, taking into account the primary interest of the marine environment in the context of the present report.

a. Metabolism

4 principal factors influence the toxic concentration at the level of active receptors:

a.1 Absorption

This requires the transfer of molecules across various membranes. The transfer can occur by four mechanisms:

- simple diffusion (by a concentration gradient),
- filtration (eg. small hydrophilic molecules),
- pynocytosis (change of the cell membrane),
- active transport associated substances.

For the greater part of these materials, transfer is by simple diffusion.

Factors influencing gastro-intestinal absorption:

- dissociation constant and lipo-solubilization of non-dissociated forms,
- stability and solubility as a function of pH,
- nature of the solvent,
- rate of solubilization,
- possible chelation.

A cellular membrane can be compared to a lipoidic membrane.

a.2 Distribution

Certain chemical substances concentrate preferentially in certain tissues. This localization is not necessarily its site of primary activity. This distribution can affect various areas (tissues, proteins). A general rule is that association is with albumen and is reversible.

a.3 Metabolic transformations

These transformations occur more readily with polar compounds, and thus are more easily excreted. These transformations are mainly catalysed by enzymes of the liver (endo-plasmic reticulum).

Besides, the possible role of intestinal bacteria should be noted.

1. Principal reactions

- enzyme oxido-reduction,
- hydrolysis,
- conjugation,
- various reactions...

2. The consequences of these transformations

- help excretion,
- produce, sometimes, a more toxic substance,
- more frequently to reduce toxicity (eg. cyanides → thiocyanates).

3. Factors affecting metabolism

- genetic,
- physiological (age, sex, hormone, illness),
- environmental : stress (cf. potentialisation).

a.4 Excretion

Several pathways.

1. Urinary: by three processes

- glomerular filtration,
- passive tubular transport,
- active tubular transport.

→ There is a tendency to produce metabolites less liposoluble and more polar.

2. Biliary

- for substance of high PM,
- for conjugated compounds (entero-hepatic cycle),

3. Gastro-intestinal

4. Perspiration, saliva, milk: small importance.

b. Interactions between compounds

In liquid medium, as in the industrial context (professional exposures), the exposure to chemical compounds is multifold; it is thus possible that interactions between compounds will occur and bring about:

. an increase (synergy, potentialisation) or,

. a decrease (neutralization) of the toxic effects.

2 types of mechanisms:

b.1 Physico-chemical interactions

- complexation,
- chelation.

b.2 Biological interactions

- at the primary reaction sites:
  . synergetic reaction,
  . additive reaction,
  . neutralization (the most frequent):
- with de-toxication mechanisms:
  . diminution,
  . increase in toxicity.

c. Toxicity

c.1 Form of toxicity: 3 types

- **acute**: exposure time short and absorption of toxic substance rapid,
- **sub-acute**: repeated exposure in a short period of time,
- **chronic**: repeated exposure over a long period of time.

Resulting pathological effects are:

. either by accumulation of the toxic substances (Q eliminated < Q absorbed),

. or by additive effects.

N.B. A third possibility, intermediary, can occur: accumulation of toxic material, but the appearance of the toxicity occurs only after its mobilization.

c.2 Types of action

- **local:** action at the point of contact,

- **systematic:** action at a distance.

Factors responsible for elective action:

. degree of perfusion of the organ,

. chemical composition of the organ,

. position with regard to transport route,

. bio-chemical characteristics:
  - important mitosis,
  - metabolization into a more toxic substance,
  - special metabolic need.

c.3 Factors influencing the response of the organism

The strength of the response depends upon the concentration of the toxic compound and the length of time of its presence. The response is governed by:

- **the physico-chemical properties** of the toxic substance, solubility, reactivity, particle size,

- **experimental factors:** absorption routes, solvent...,

- **biological factors:** metabolism, reactivity of receptor organs,

- **environment:** temperature, humidity, interactions.

c.4 Entry routes into the organism

Out of the five principal routes (oral, pulmonary, cutaneous, occular and parenteral) only the digestive route is of importance here.

Not taking into account local effects (direct effects on the digestive tract), the response (reaction) depends above all on the degree of absorption: the liver is the most important (target-organ) (absorption by the lymphatic system via the thoracic canal is of little importance even for substances which are very lipo-soluble).

N.B. For cutaneous and occular pathways, four factors determine the absorption:

- **physico-chemical properties of the compound,**

  . penetration,

    . by sedaceous cells,

    . by epidermic cells.

. solubility in water or in lipids,

. ionization and molecular weight.

- solvent used,

- integrity of the skin.

Example of sensibilysing elements: nickel, mercury, chromium^{VI}.

c.5 Evaluation of toxicity

1) **Determination of lethal dose 50** (LD50) is needed in order to have some measure of the intrinsic toxicities of different compounds. These determinations will be insufficient in the context of pollution prevention.

2) Only a multi-parameters experiment aimed at determining **the reaction mechanism and the metabolic function** of the polluting compound will give sufficient information on the sub-lethal concentration (standard tests are of little value).

c.6 Toxic reaction mechanisms

1) The primary action of toxic substances on cellular or extra cellular material:

- **interference** with oxygen transfer processes (action on haemoglobin) - cf. CN^-, enzymes and cytochrome oxidase,

- **action on enzymes**

  . **inhibition**

  this can be direct or indirect
  . interference with activator or co-factor,
  . inhibition of the synthesis of an enzyme,
  . conversion of toxic derivatives,

  . **stimulation** (of microscopic enzymes),

- **hypersensitivity:** the toxicity results from the inactivity or absence of an enzyme,

- **direct action on a receptor.**

2) Secondary actions

Additional biological modifications.

6.10 IMPACT OF MARINE POLLUTION ON MAN

The impact of marine pollution on man follows three possible routes:

- direct pathological injury,
- poisoning of food sources,
- loss of the ability to exploit or enjoy marine areas due to pollution.

In this context, the role of metallic compounds (and especially those under study in the present report) must be understood, not only in the light of their specific reactions, but also in their concerted actions.

a. At the level of contact

An exceptional risk, but one nevertheless of importance, is that of the proliferation of bacteria in certain areas in which the self-purifying ability of the sea water has been greatly diminished due to excessive input of pollutants (neo-pathological exposure in seaside resorts). This problem, although not specific to metal compounds, may result from a restricting action of the materials on the biocoenotic equilibrium.

b. At the level of food supply poisoning

The disease of Itai-Itai (whatever the real pathogenic importance of cadmium) is a sufficient example of this risk.

These types of accidents will occur more often if we simply argue on the exceptional circumstances. Prior to measurements, one should test the ingestibility of proposed food products (certain precautions taken in oyster-culturing should not mask the near total lack of regulations or practice along these lines).

Elsewhere, the introduction of certain technologies in tropical marine areas may, at least in part, be held responsible for the phenomena of ichytyotoxism (eg. Ciguatera). This is the more worrying because it concerns food sources of populations largely dependent on the sea for its staple diet.

Finally, a case has been recorded where the desalination of sea water has led to the initial symptons being discovered of a disease caused by the consumption of drinking water containing too high concentrations of dissolved salts (for example zinc).

c. Degradation of the environment

The waste released by raw material using industries, characterized by heavy metals, causes an undeniable degradation of continental waters (especially of river and outlet waters).

However, its impact on the bios cannot be separated from the total effect caused by other stresses (cumulative effects, synergistic, etc.). Attempts at preservation (in certain regions, one speaks of restoration) need to take into account all sources of pollutants.

Only such an approach will stop the transformation of certain coastal waters into marine deserts with all the attendant consequences on man's coastal activities.

6.11 RESEARCH ON THE PRACTICAL THRESHOLD OF THE MARINE ENVIRONMENT

From consideration of all the variables examined above, it can be clearly seen that for a given pollutant (eg. metal salt) introduced into the marine environment, there is no single specific action on the bios, for a given concentration, but rather a multitude of reactions from which one can infer that all bio-stress caused by a given toxic compound leads to a particular degradation of the environment at any particular place.

In other words, there does not exist, at the present time, any generalized model of analysis or of control, which allows an overall understanding of the real toxic impact of any given pollutant.

All efforts to codify these problems are made less effective because of the complexity and natural instability of the living environment, based on its adaptive ability. Because of this, it is perhaps of interest to develop a biological code in order to have a better understanding of biological phenomena, where very often technologism tries to impose its own language.

N.B. The most important problems which refer to the rights and health of consumers (and users) should be based upon the principle of the integrity of all exploitable resources; this concerns equally living resources and more importantly potential resources rather than those already under exploitation (this is to ensure that instead of just preserving marine bio-productivity for fishing uses for example, that wider exploitation within other sectors is preserved; agro-chemical, pharmaceutical, bio-medical, etc.).

Besides, and in face of the undeniable and worrying degradation of the marine environment, it is necessary to give urgent priority to measures which will prevent (and so minimize) it by an adapted prevention technology by reducing the causes (waste materials) using the appropriate technical steps necessary. This is, above all, an industrial problem which industry is asked to resolve by the application of techniques of which some are already in use.

CHAPTER 7

CONCLUSIONS
GENERAL AND SPECIFIC RECOMMENDATIONS

CHAPTER 7 - CONCLUSIONS, GENERAL AND SPECIFIC RECOMMENDATIONS

In the face of the undeniable degradation of the marine environment, the present attempt at appreciating the real impact of various pollutants illustrates the importance of the following three points:

- the complexity of the response by the bios to all types of stress,
- the concerted influence of various types of pollutants,
- the necessity to initially regard the problem from each local region concerned.

General interest, the importance and diversity of the measures to be taken, demand an in-depth analysis, especially for the marine environment, of the dispersion of certain polluting elements resulting from industrial operations. This analysis should lead to the formation of an inventory of prevention techniques.

A certain number of general and specific recommendations have been formulated and are presented below. These recommendations take into account the actual state-of-the-art in understanding, the techniques available or those in the course of development, as well as their economic implications.

However, the problem of marine pollution should not be considered as singly caused by production industries, but equally as a result of all urban concentrations, of consumption and of raw discharge of waste.

7.1 GENERAL RECOMMENDATIONS

a. Behaviour of wastes (salts and compounds) in the marine environment

Concerning, in general, metal compounds in certain industrial effluents, much work has been carried out by scientific laboratories on the fate of these compounds, which, in the marine environment, may be transformed firstly by the physical and chemical properties of the environment (pH, t°, $s/_{oo}$, dissolved oxygen, etc.) and subsequently by biological phenomena (ingestion-fixation, metabolization, concentration, degradation, etc.).

The modes of transfer and transformation in sea water are of great importance, and it is proposed that an inventory of data in this area should be gathered on those elements that are of primary concern to us; *chromium, zinc, cadmium,* and based on these data to propose regions of complementary work, especially those areas concerning physico-chemical forms.

b. The problem of reproducibility of analytical results

In a liquid medium, the sea in particular, when the level of concentrations is extremely low*, sampling methods, transport and conservation treatments, and analysis of samples must be considered with particular care if a high degree of confidence in the results is to be achieved.

* *for example < 10 µg/l*

The current methods of analysis are often inadequate due to the very low concentrations encountered, and above all to the fact that a part of the metals can be adsorbed on particles and sediments or taken up by the marine bios.

In this framework it is recommended that an intercomparison of methods and above all of results be carried out, both within and between laboratories, in order to assess the value of data, particularly data published on this subject in recent years.

This problem is of special importance in the field of detection and analysis of micro-trace metallic elements in *biological mediums* and it is thought indispensable that a number of European inter-laboratory comparisons be organized in order to determine levels of reproducibility. We recommend that a similar approach should be adapted in the case of *sea water*.

c. Elsewhere, the problem of pollution arising from the final consumption of certain products and objects based on heavy metals must be evaluated. In effect certain preventative measures, at the production level, may be considered insignificant in comparison to the quantities of metals reaching the environment, as wastes, via intermediate and ultimate product consumption. This point particularly requires special attention. The solution of this problem is complex and may eventually depend upon certain contingent uses of the heavy metals in question.

d. Finally, the question of essorage of waste muds from all types of treatment installations must be considered. One of the members of the European Oceanic Association who participated in the present study and who possesses wide experience in this field, would be willing to continue, at the request of the Commission of European Communities, and to develop research plans in order to resolve this essential question.

7.2 SPECIFIC RECOMMENDATIONS

a. Preliminary Considerations

- A summary of preventative techniques at present shows that all are based on the prevention of polluting elements in the final effluent. No attempt is made here to discuss possible changes which may be caused by new production processes, completely integrated systems, of engineering conceptions and of measures destined to ensure completely clean production.

  On the other hand, the importance of those elements discussed in the present report should be underlined, with particular emphasis on preventative techniques which allow the elimination of the products concerned by allowing their recovery and/or recycling at a good economic level.

It is possible for a prevention technique to be efficient if pollution threshold limits are attained, but as far as the economy of the operation is concerned, taking into account investment and running costs, important difficulties may remain. Thus, *recovery and recycling* of elements becomes a necessity, depending upon the actual economy of the resources.

The profitability of prevention operations, in particular by recycling of products, requires the introduction of the notion of *critical size* (i.e. the size of the polluting sources concerned). This idea applies especially to industries consuming small amounts of water.

. In consequence, it is recommended, as much for those industries situated on or near the coast as for those inland, the collection of effluents with a view to carrying out purification treatment on *centralized treatment platforms*, as is now being started in certain European countries and would appear very desirable at community level.

In particular, the promotion of secondary treatment platforms for final residues or for waste coming from primary treatment platforms, where the quantity of specialized waste is insufficient for the economic treatment at regional or national levels but could well be envisaged at international and/or community operations. A good example of this would concern the recovery of certain used (spent) catalysts, (catalysts of chromium/zinc used in various industries).

. It should be remembered that a series of control zones need to be installed for the continuous monitoring of effluents in coastal and estuarine areas.

A normalization (systemization) of control systems and analysis techniques appears to be indispensable.

b. Specific recommendations

b.1 Chromium

It should be recalled that solvent extraction of chromium can lead to the recovery of Cr^{VI} at a level strongly conforming to those considerations mentioned above. (liquid/liquid extraction for example using tributylphosphate for Cr^{VI} in effluents containing sufficiently high concentrations. Elution being carried out using chromate solution. Chromium being recovered as bichromate economically).

However, liquid/liquid extraction can leave a residual concentration of Cr^{VI} in the effluent which is greater than the admissible limits. This imposes a more rigorous final treatment.

This treatment often consists of reducing Cr^{VI} to Cr^{III} and then precipitating the latter, although in a form extremely difficult to recover.

It is also possible to fix Cr^{VI} on resins and to recover chromium in a form more easily re-usable (as chromium sulphate, elution by SO_2 in sulphuric acid medium). This procedure, however, requires a relatively large investment of money.

When the size of a projected purification installation for chromium containing effluents does not reach a certain critical size, it is recommended that a *centralized treatment platform* treats all effluents collected in a given region. A scheme for such a platform is given in figure 7.2.b.

It is believed necessary to underline here that such platforms allow not only the possibility of purified liquid effluents but also stops the transfer of pollution. In effect, the solids can be recycled on the spot, (when, for example, a platform is connected to a factory producing chromium compounds) or be sent back to the nearest production factory. In addition, platforms having a certain critical size can obviously be given adequate means of control concerning the atmospheric pollution.

For a treatment capacity corresponding to between 500 and 1,000 tons of recovered products per year, the necessary construction investment of such a platform would be between 1.5 and 2 million units of account. The running cost of the installation being in the order of 0.4 million units of account per year.

On the other hand, raw waste effluents may converge on collective water sewage treatment plants where chromium, in general, has an inhibiting effect on the functioning of bacterial beds or activated muds; it should be carefully examined how these problems may be approached by the application of regulations protecting the environment.

However, it would be possible to use certain microorganisms in order to treat effluents having as a base ions or salts of chromium. One of the participants of this report could, at the request of the Commission of European Communities, undertake, on the basis of certain existing research, a study of biological methods which could be used for the elimination of certain heavy metals such as chromium, zinc and cadmium. A proposal along these lines could be prepared rapidly.

b.2 Zinc-cadmium

- Concerning more specifically the rayon producing industries, simple precipitation treatment leads to effluent residuals containing only a few mg/l Zn*.

* *for example 3 mg/l*

FIGURE 7.2.B - PURIFICATION OF CHROMIUM EFFLUENTS

Projected centralized station associated with a production factory for chromium derivatives*

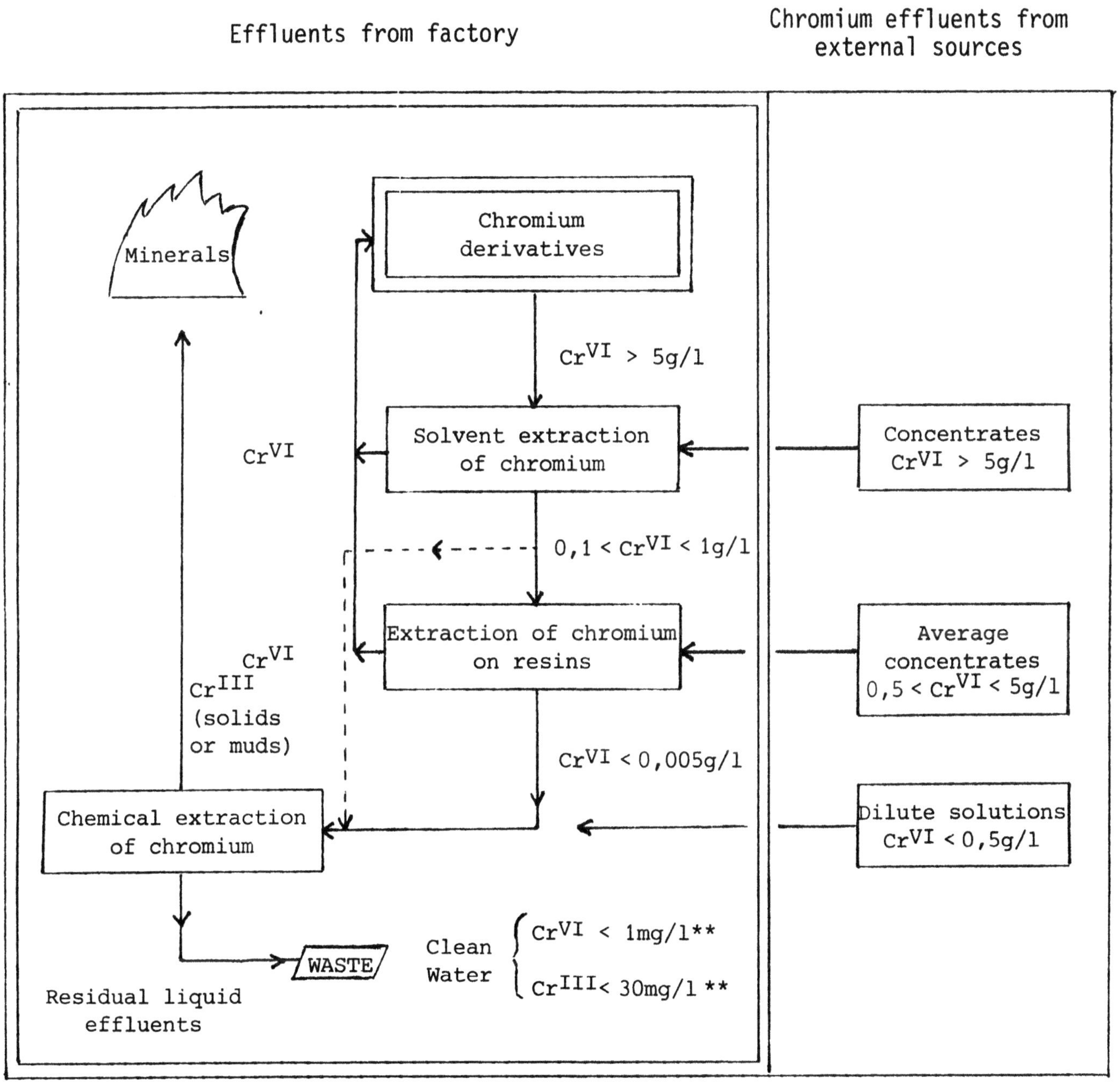

* *Private communication*

** *These grades to date can be greatly reduced in order to satisfy future changes in regulations. For* Cr^{VI}*, the residual concentration of purified waters can normally lead to a limit of less than 0.1 mg/l and a concentration of total chromium of 1 mg/l.*

- A more complete treatment in two stages - precipitation and ion exchange - should result in a final concentration of very much less than the ppm level. The actual efficiency of the process might well be influenced by the presence in treated waters of various compounds such as wetting agents, complexing agents and other materials. Results so far obtained by laboratory experiments on such a method suggest the need to determine more precisely the residual concentrations of metals in the effluents, as a function of various compounds added at different stages of production. An understanding of these quantitative relationships should allow a better estimation of the limits of application of this promising technique for the recycling of zinc.

  In a more general manner, for chemical purification units for Zn/Cd and even Cu/Ni effluents, treatment would consist of adjustment of pH and addition of chalk, such that complete precipitation of metallic hydroxide is assured. The last traces of non-precipitated metals can be precipitated as sulphides. Attention should be given to the danger of residual pollution.

  Elsewhere, purification or recovery of effluents by electrolysis, whether combined with ion exchange treatment or not, should be made the object of feasibility studies, principally for those metallic elements within the scope of the present report, zinc and cadmium, and possibly for other heavy metals such as chromium or lead. The results already obtained on pilot plants are sufficiently encouraging to look in greater depth at the technical and economic possibilities of this method.

- Concerning effluents containing cadmium, as for example those coming from cadmium electroplating processes (workshops) not being sufficiently large for the installation of efficient purification plants, the preceding recommendations made for chromium containing effluents (centralized platforms for purification processes) are equally viable for cadmium.

- It is to be noted that, in general, rainwater of low pH (i.e. acidic) dissolves, in certain regions, relatively high amounts of zinc and also, eventually, allied metals from metal structures standing in the open. It would seem to be of interest, in such regions, to know the corresponding levels of pollution, and taking these into account, prepare plans for the prevention of pollution, both up and down stream.

b.3 Cyano-metal compounds

There exist numerous methods of destruction of cyano-compounds present in various industrial effluents.

The following points should be noted:

- classic treatment with hypochlorite, which now tends to be overtaken when:

- in difficult cases, treatment with caro acid. This gives excellent results but at a relatively high cost,

- more frequently, treatment with hydrogen peroxide, this method is particularly useful and will become probably in the future the routine method of treatment,

- catalytic oxidation process (cyancat). This is based on a totally different method but will, without doubt, be used increasingly often, especially in centralized purification platforms,

- lastly, biological treatment using certain types of micro-organisms. The procedure is simple, anaerobic, the active micro-organisms growing on special plastic support material.

It should be mentioned that a large number of electroplating factories, relatively modest in size, are not yet equipped to purify their effluents or have financial difficulties in installing such equipment. It is strongly recommended that such factories should send their wastes to centralized purification platforms of a type already discussed in the context of chromium or cadmium containing effluents, these can also be of multi-element purification design.

However, these factories should be equipped with non-diluted baths, either by modifications to rinsing circuits or by changes to general drainage circuits.

7.3 SUGGESTIONS FOR PILOT ACTION PLANS

a. Multi-element centralized purification platforms

Certain members of the European Oceanic Association participating in the present study have elaborated the essential outlines of a plan of recommendations which consists of:

- a comparative enquiry into existing, or projected, centralized treatment installations,

- the definition of adequate centralized platforms for certain numbers of regions within the Community,

- the evaluation of the impact of such platforms on the reduction of pollution, principally that coming from small to medium enterprises,

- conditions of economic functioning (provisional study of profitability).

This proposition, to be carried out over one year, could be given in more detail at the request of the Commission of the European Communities.

b. **Establishment of a dispersion and accumulation model for a given region in marine environment for those polluting elements studied in the present report (chromium, cadmium, zinc, cyano-metal compounds).**

Due to the importance of this problem, if applied to an industrial region of the Mediterranean, a study time of three years would be required. This is based on a proposition outlined by one of the members of the European Oceanic Association, who took part in the present study.

The principal stage of such a study would comprise among others:

- the development of an integral schematic model, including physical, chemical and biological model inputs,

- the experimental determination of corresponding data in the region under consideration,

- a study of the correlation between the developed model and the experimental data obtained.

Here again, on the request of the Commission of European Communities, a detailed proposal could be made.

It is probably useful to add at this point that the possibility of obtaining a generalized model for the dispersion of pollutants in the marine environment has been recommended and discussed at the Rome Conference of December 1970. This conference was organized by the FAO and was on the general theme of marine pollution and its effects on living resources. It was recognized that a certain number of difficulties, at that time, rendered this objective beyond the bounds of possibility, except in the case of excessive simplification of the model.

In effect, even if the mathematical model approach concerning each of these factors: physical behaviour, chemical alterations and biological fate seemed possible, the number of factors and their possible combinations were so large that computers of sufficient size were still not available.

c. Taking into consideration the interdependence of these factors such that the prediction capability of a model is limited by the understanding of the least known variable, it would be, perhaps, interesting to suggest a research programme aimed at determining the weakest element (the least known variable) of the system.

Under this heading, it is proposed that a study of the sorption of polluting elements (heavy metals and/or organo halogens) by marine sediments represents a subject of interest that should be looked at in depth. A proposition along these lines is under study by one of the members of the European Oceanic Association.

ANNEXES

NOTE

The tables shown in the two following annexes give the general synoptic diagrams of pollution resulting from the production and various utilizations of chromium, cadmium and zinc. It is envisaged to complete these tables by a series of indexes giving the different production tonnages and losses as and when the corresponding information is obtained.

A N N E X E 1

GENERAL SYNOPTIC DIAGRAM OF POLLUTION
CAUSED BY THE PRODUCTION AND THE VARIOUS USES
OF CHROMIUM

GENERAL SYNOPTIC DIAGRAM OF POLLUTION CAUSED BY THE PRODUCTION AND VARIOUS USES OF CHROMIUM

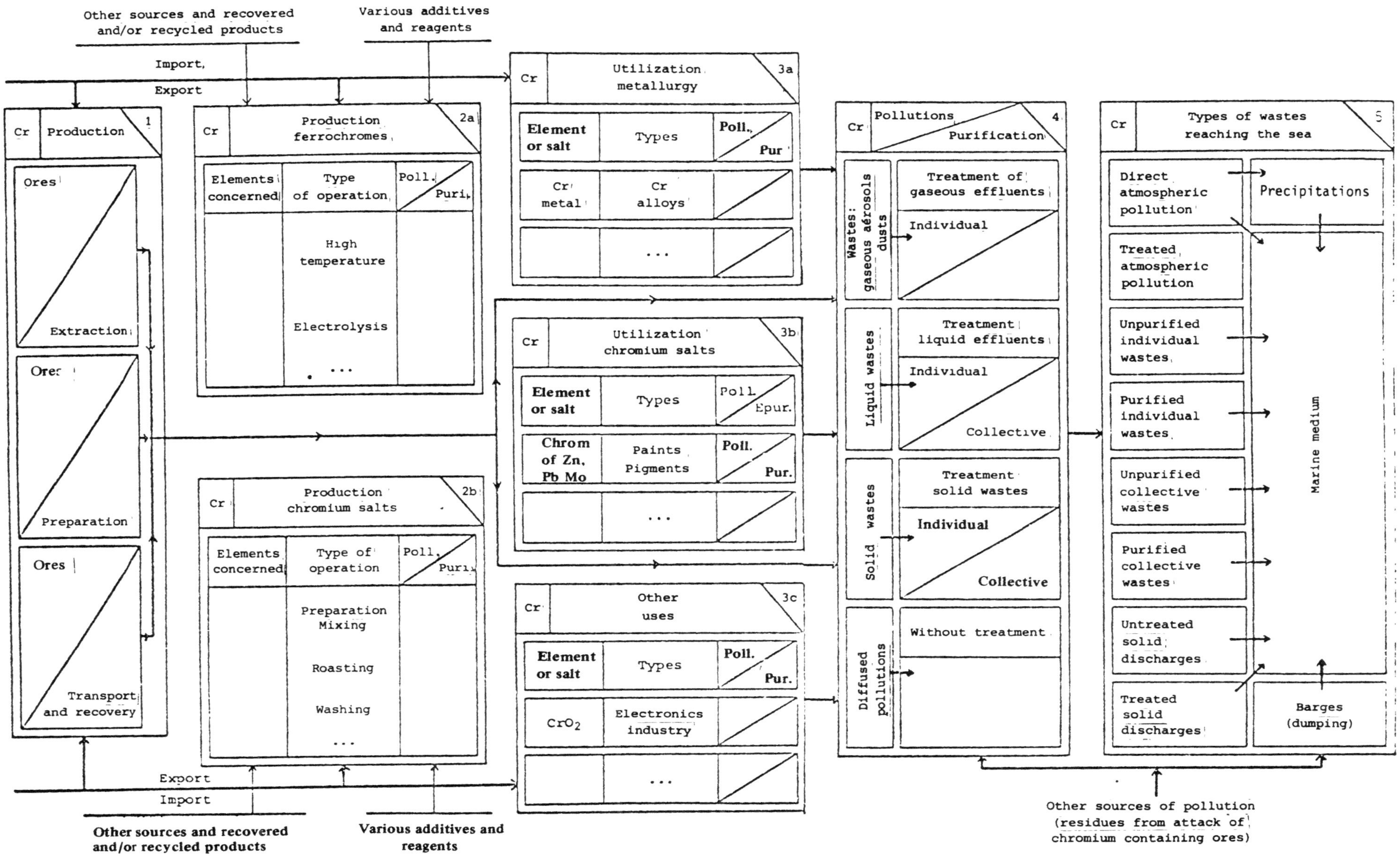

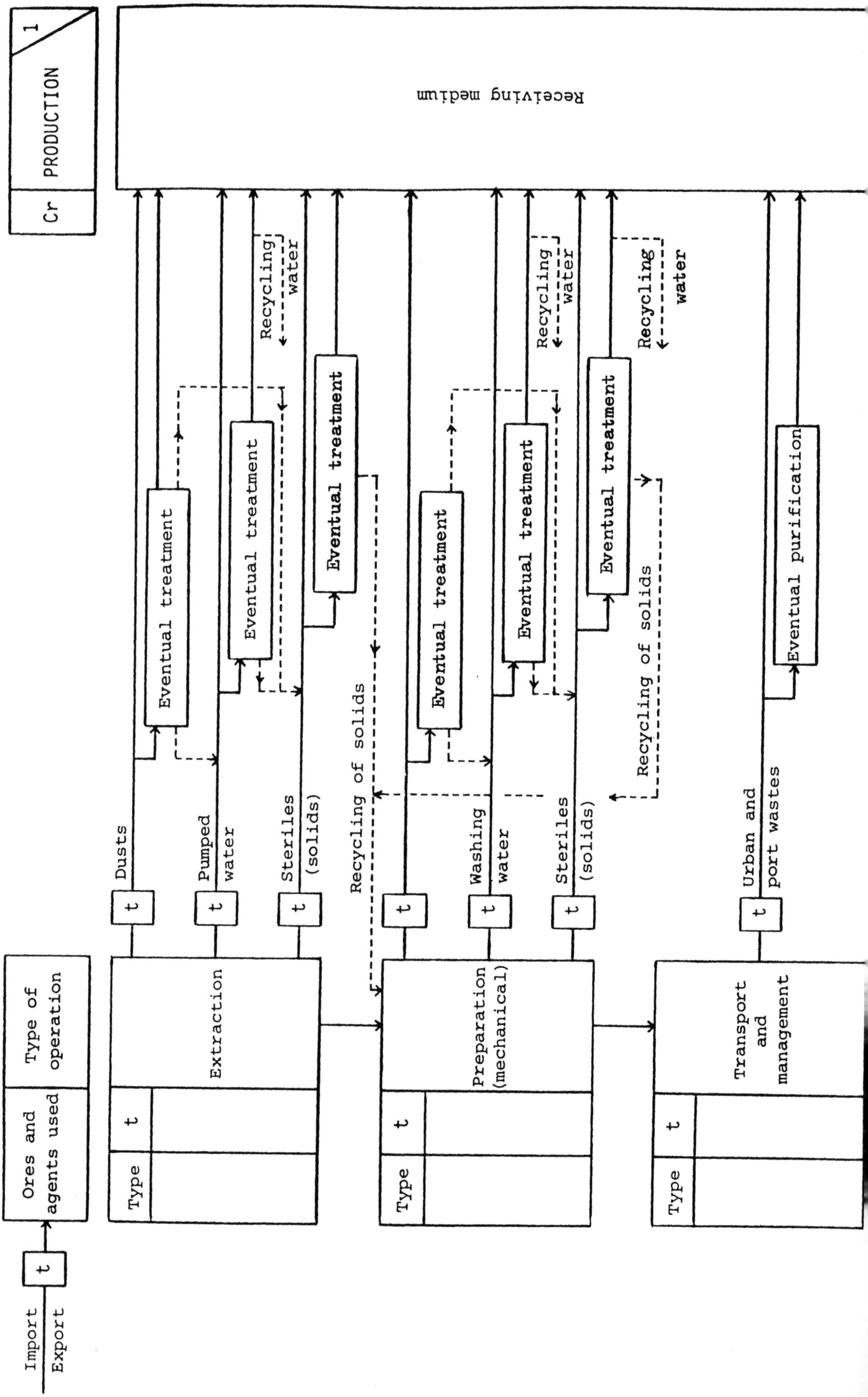
Cr
PRODUCTION
1
Receiving medium
Import
Export
t
Ores and agents used
Type of operation
Extraction
Preparation (mechanical)
Transport and management
t
Type
Dusts
Pumped water
Steriles (solids)
Washing water
Urban and port wastes
Eventual treatment
Eventual purification
Recycling water
Recycling of solids

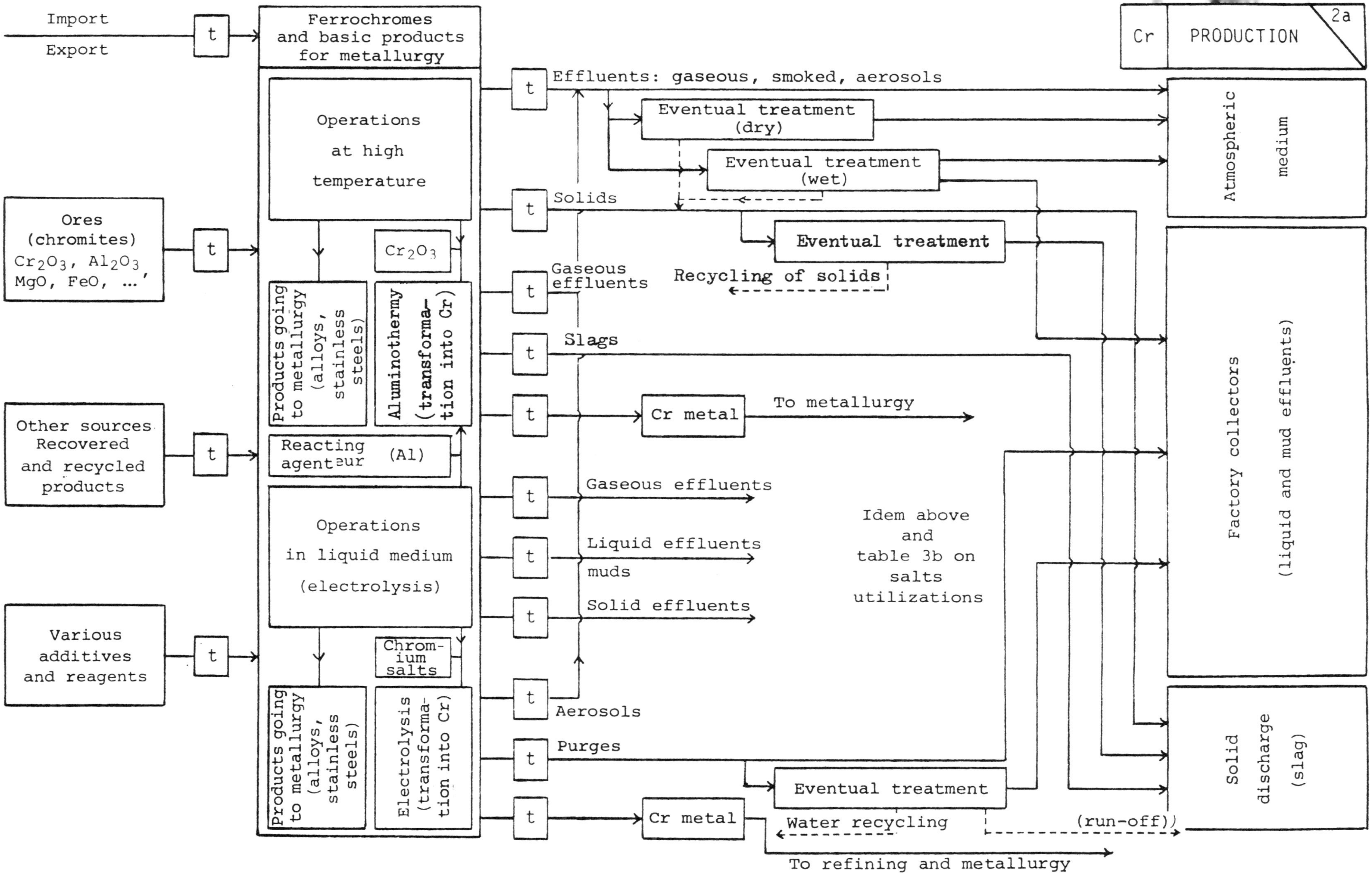
Cr
PRODUCTION
2a
Import
Export
t
Ferrochromes and basic products for metallurgy
Ores (chromites) Cr_2O_3, Al_2O_3, MgO, FeO, ...,
Other sources Recovered and recycled products
Various additives and reagents
Operations at high temperature
Cr_2O_3
Products going to metallurgy (alloys, stainless steels)
Aluminothermy (transformation into Cr)
Reacting agenteur (Al)
Operations in liquid medium (electrolysis)
Chromium salts
Electrolysis (transformation into Cr)
Effluents: gaseous, smoked, aerosols
Eventual treatment (dry)
Eventual treatment (wet)
Solids
Eventual treatment
Gaseous effluents
Recycling of solids
Slags
Cr metal
To metallurgy
Gaseous effluents
Liquid effluents
muds
Solid effluents
Idem above and table 3b on salts utilizations
Aerosols
Purges
Eventual treatment
Water recycling
(run-off)
To refining and metallurgy
Atmospheric medium
Factory collectors (liquid and mud effluents)
Solid discharge (slag)

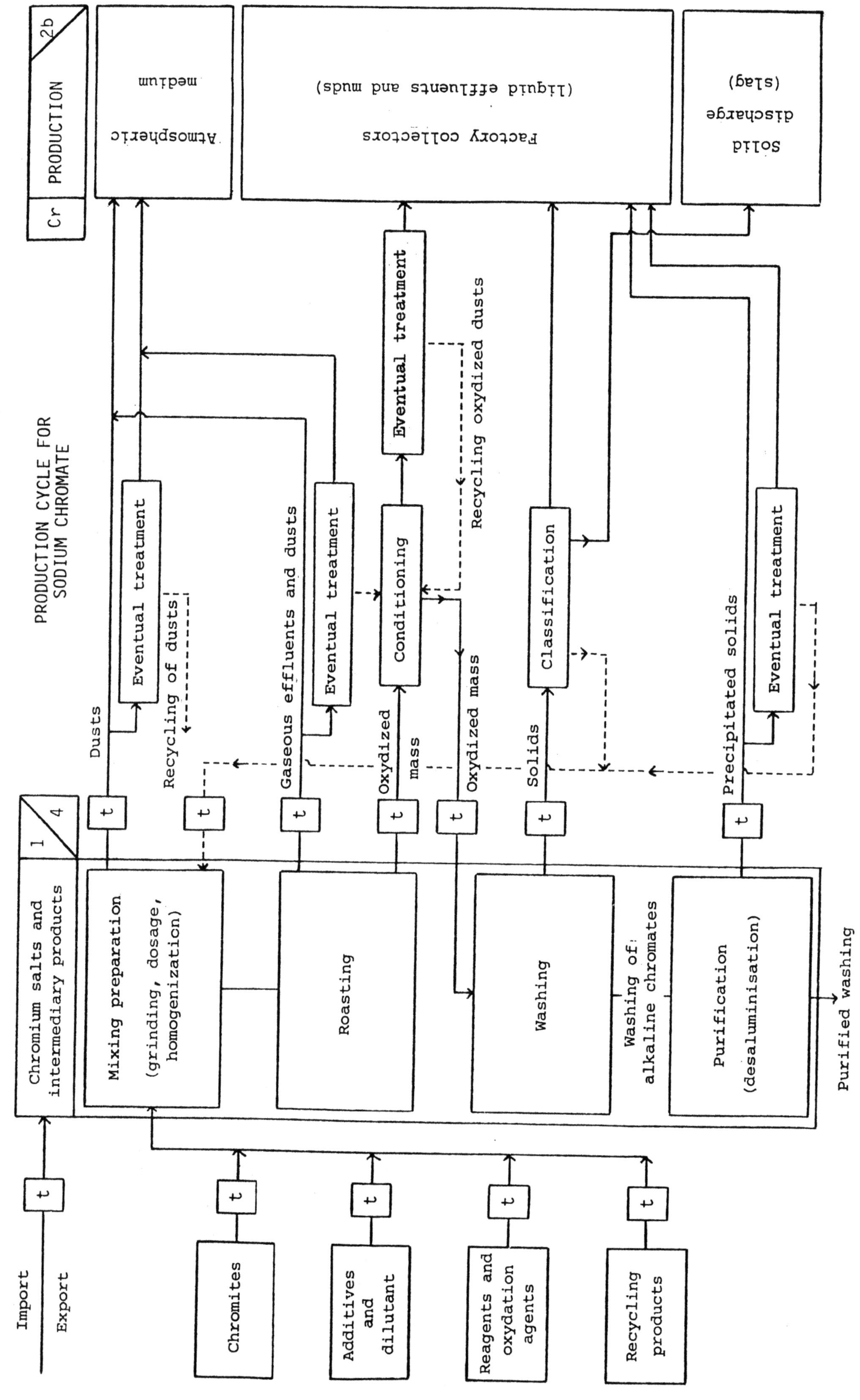
PRODUCTION CYCLE FOR SODIUM CHROMATE
Cr
PRODUCTION
2b
1
4
Chromium salts and intermediary products
Import
Export
t
Chromites
Additives and dilutant
Reagents and oxydation agents
Recycling products
Mixing preparation (grinding, dosage, homogenization)
Roasting
Washing
Washing of alkaline chromates
Purification (desaluminisation)
Purified washing
Dusts
Recycling of dusts
Gaseous effluents and dusts
Oxydized mass
Oxydized mass
Solids
Precipitated solids
Eventual treatment
Eventual treatment
Conditioning
Eventual treatment
Recycling oxydized dusts
Classification
Eventual treatment
Atmospheric medium
Factory collectors (liquid effluents and muds)
Solid discharge (slag)

PRODUCTION CYCLE FOR SODIUM BICHROMATE

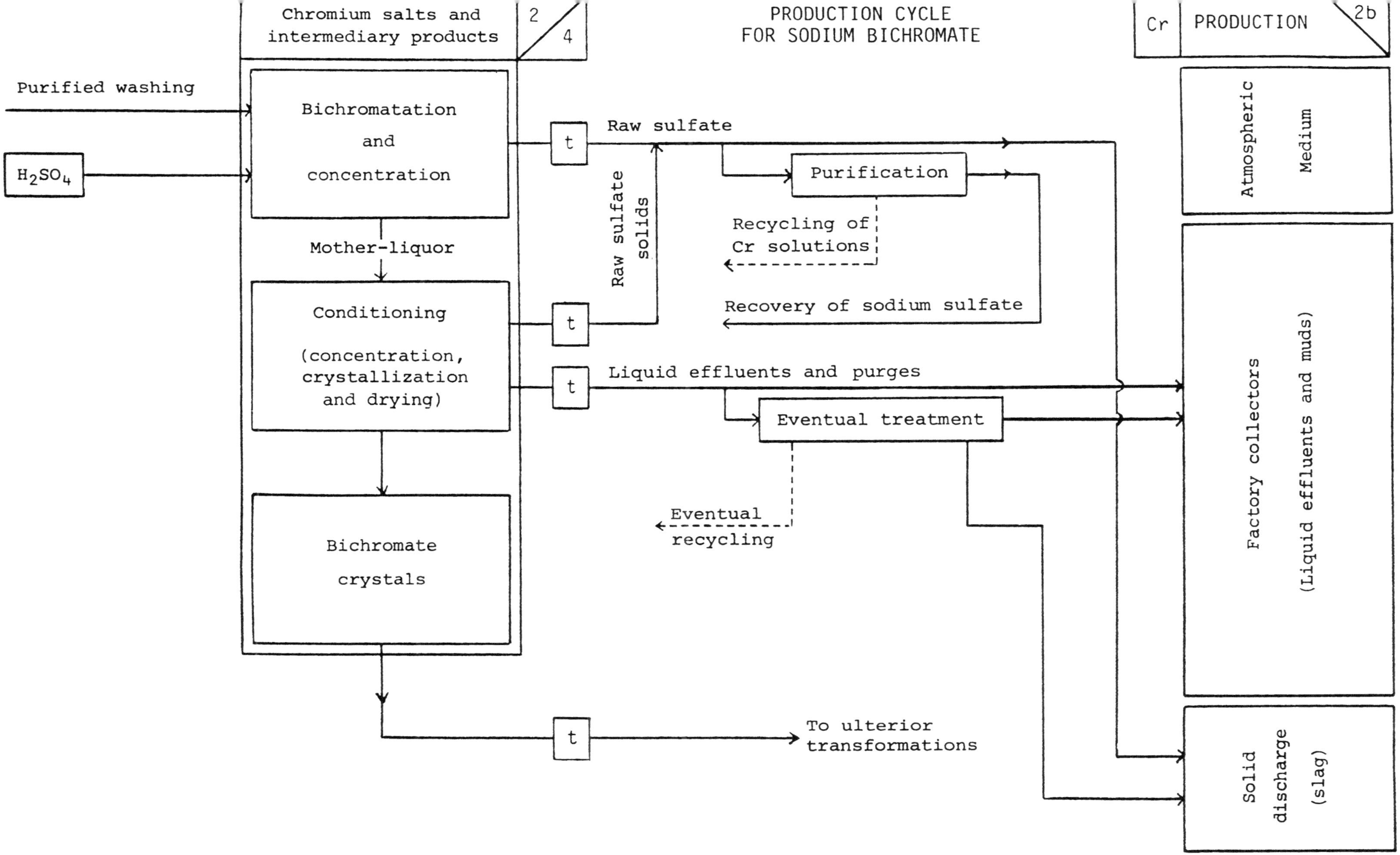

PRODUCTION CYCLE FOR SODIUM BICHROMATE DERIVATIVES (CrO_3 and Cr_2O_3)

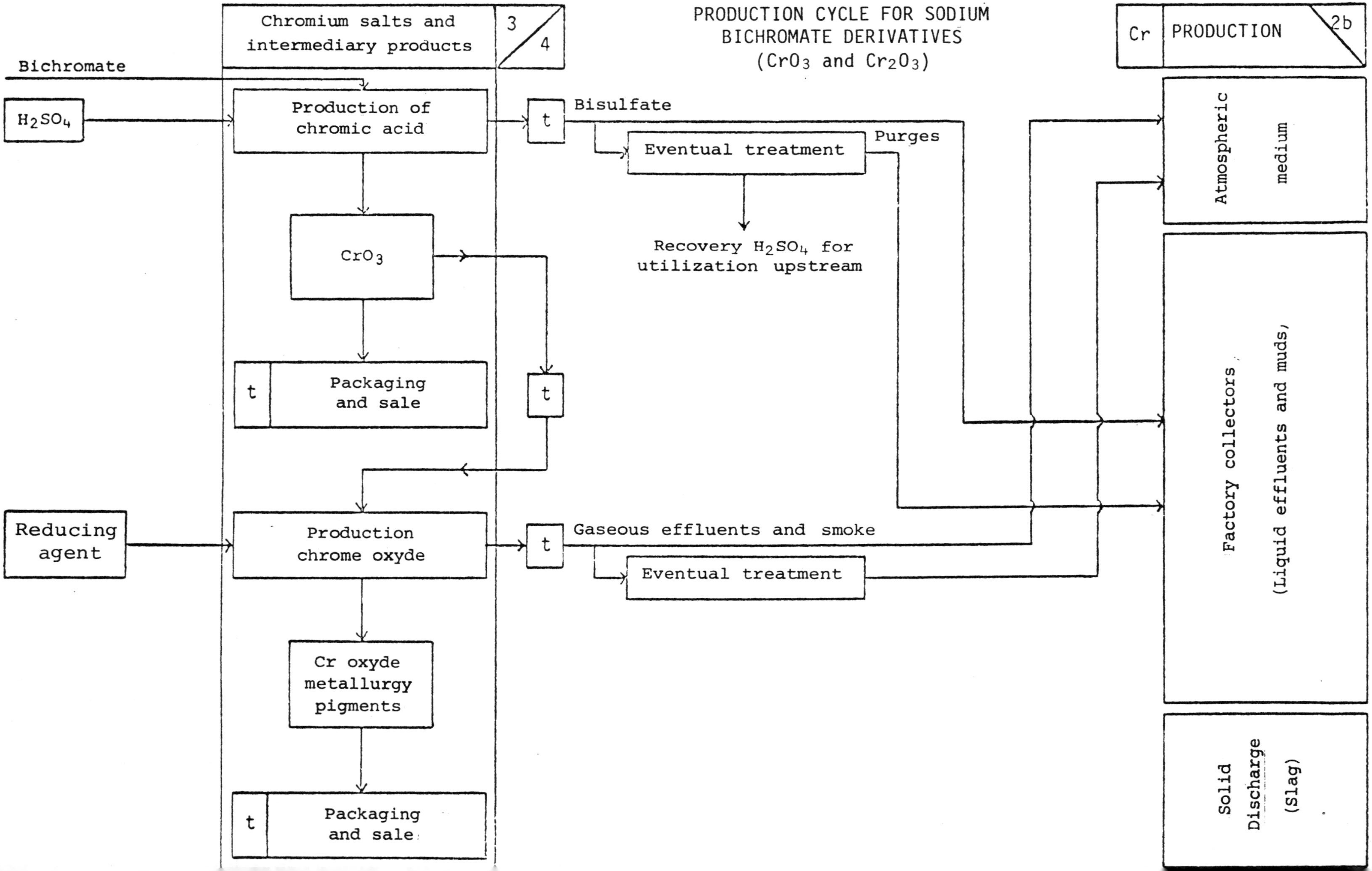

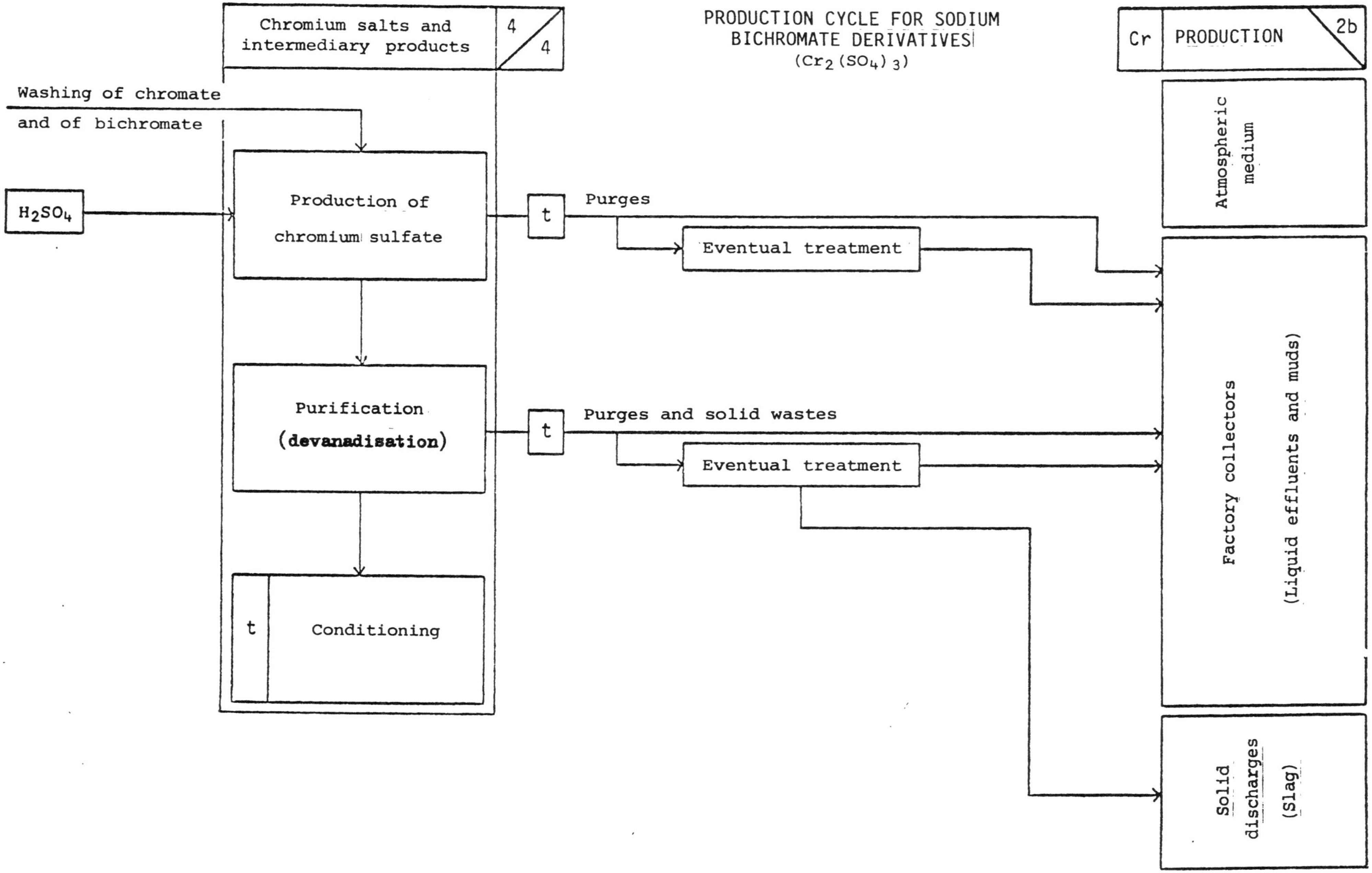
Chromium salts and
intermediary products
4
4
PRODUCTION CYCLE FOR SODIUM
BICHROMATE DERIVATIVES
($Cr_2(SO_4)_3$)
Cr
PRODUCTION
2b
Washing of chromate
and of bichromate
H_2SO_4
Production of
chromium sulfate
t
Purges
Eventual treatment
Purification
(devanadisation)
t
Purges and solid wastes
Eventual treatment
t
Conditioning
Atmospheric
medium
Factory collectors
(Liquid effluents and muds)
Solid
discharges
(Slag)

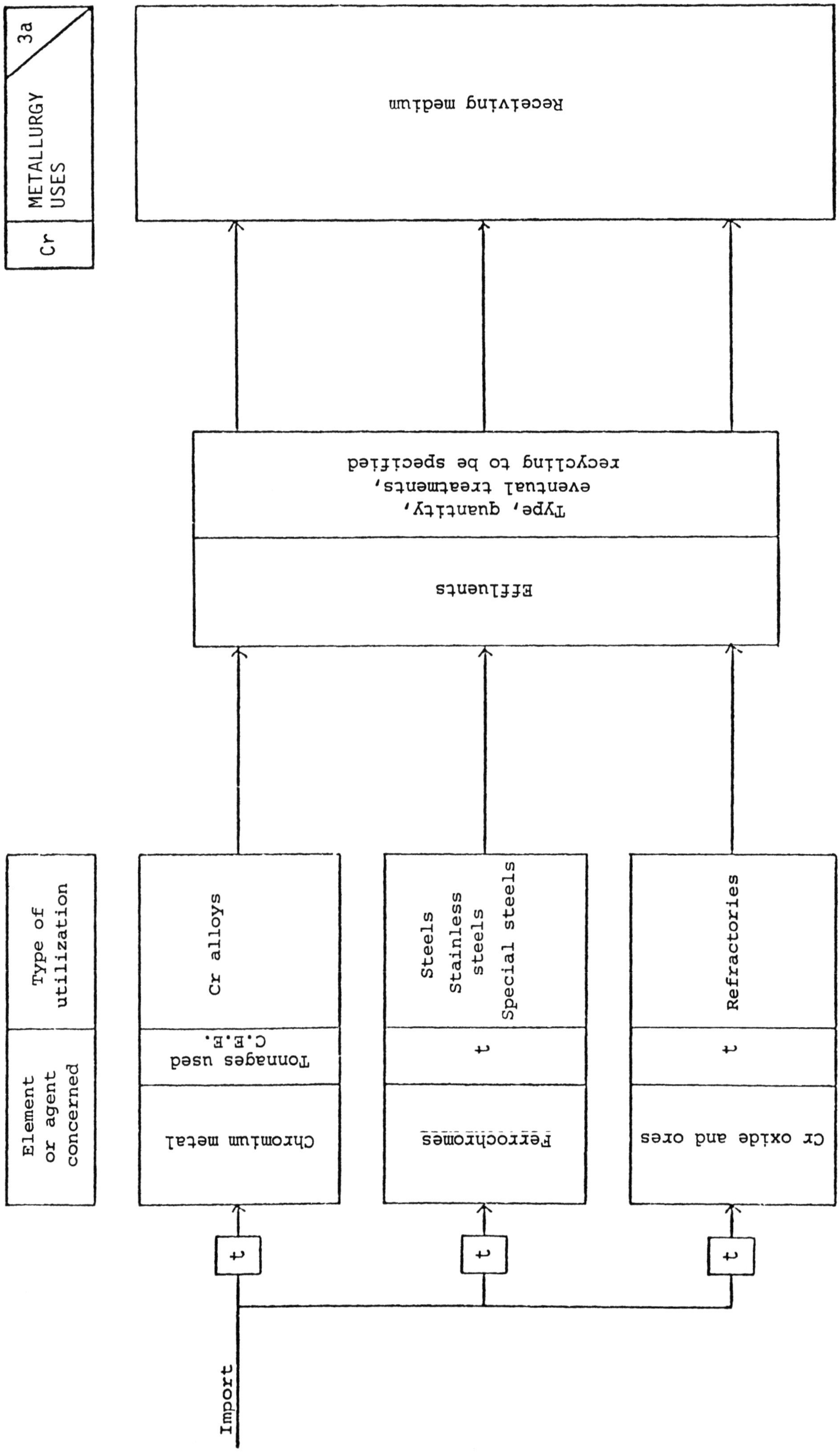

Cr
METALLURGY USES
3a
Element or agent concerned
Type of utilization
Import
t
Chromium metal
Tonnages used C.E.E.
Cr alloys
Ferrochromes
t
Steels
Stainless steels
Special steels
Cr oxide and ores
t
Refractories
Effluents
Type, quantity, eventual treatments, recycling to be specified
Receiving medium

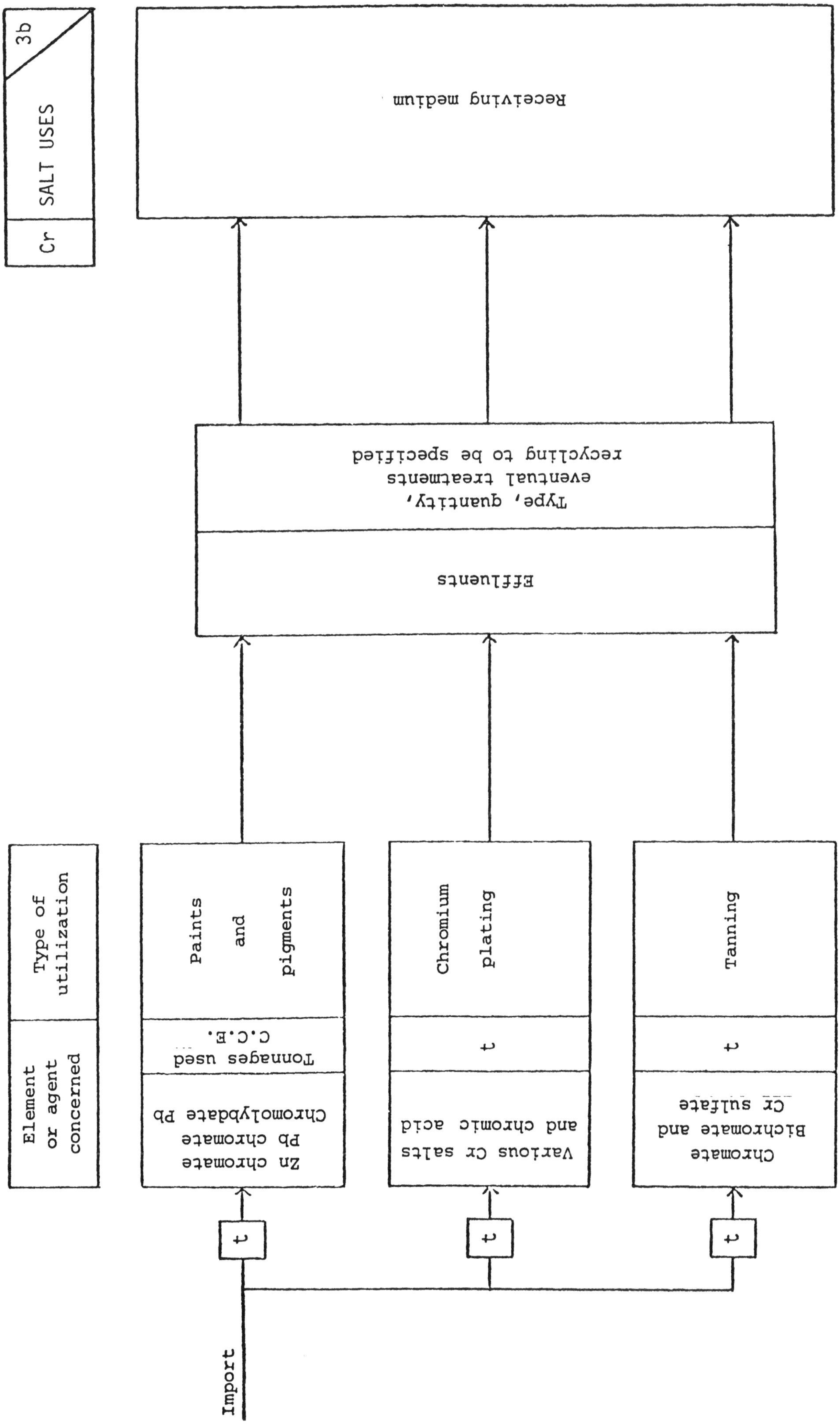
Cr
SALT USES
3b
Receiving medium
Type, quantity,
eventual treatments
recycling to be specified
Effluents
Element or agent concerned
Type of utilization
Zn chromate
Pb chromate
Chromolybdate Pb
Tonnages used
C.C.E.
Paints and pigments
Various Cr salts
and chromic acid
t
Chromium plating
Chromate
Bichromate and
Cr sulfate
t
Tanning
t
t
t
Import

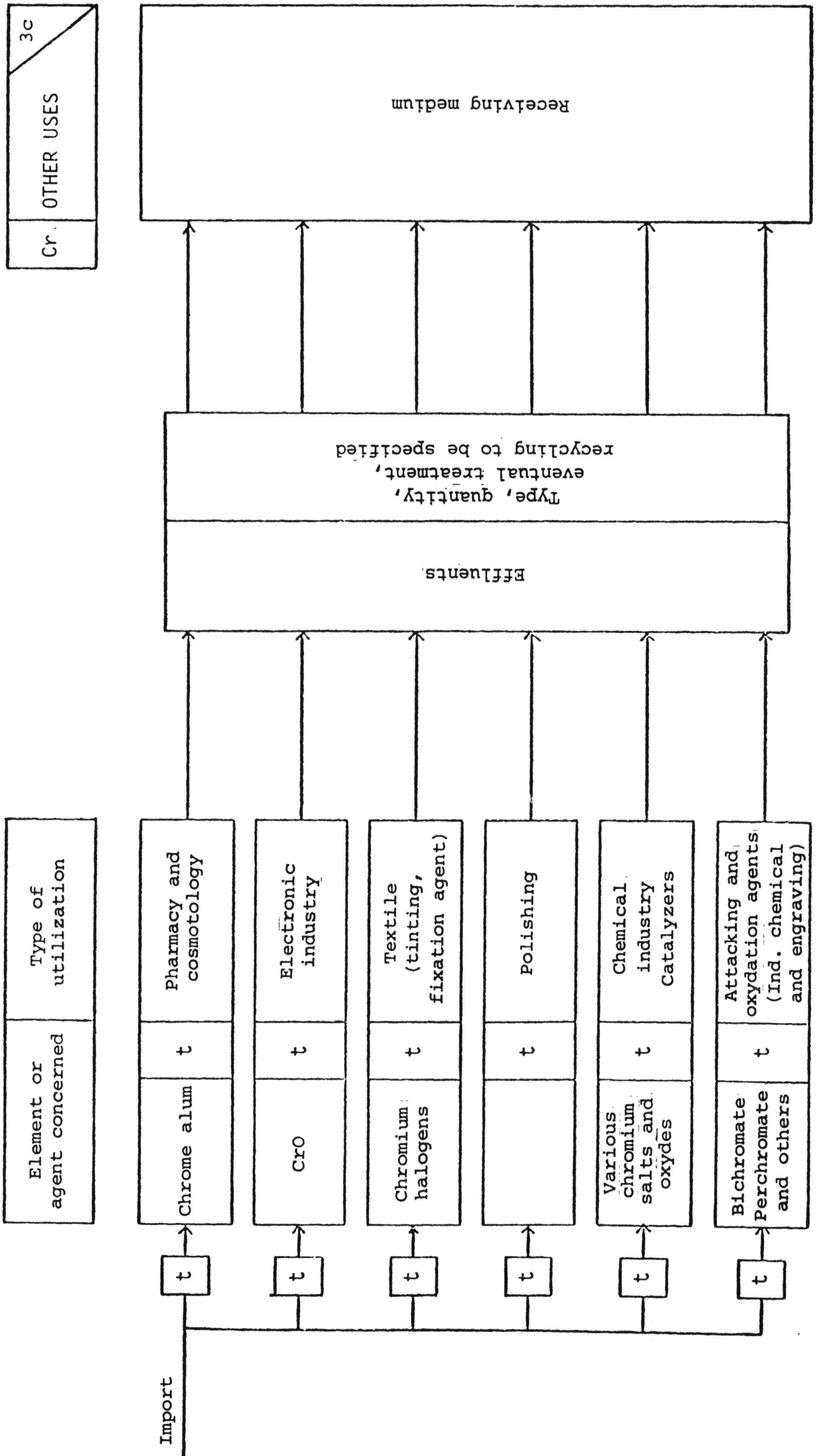
Cr
OTHER USES
3c
Element or agent concerned
Type of utilization
Chrome alum
t
Pharmacy and cosmotology
CrO
t
Electronic industry
Chromium halogens
t
Textile (tinting, fixation agent)
t
Polishing
Various chromium salts and oxydes
t
Chemical industry Catalyzers
Bichromate Perchromate and others
t
Attacking and oxydation agents (Ind. chemical and engraving)
t
Import
Effluents
Type, quantity, eventual treatment, recycling to be specified
Receiving medium

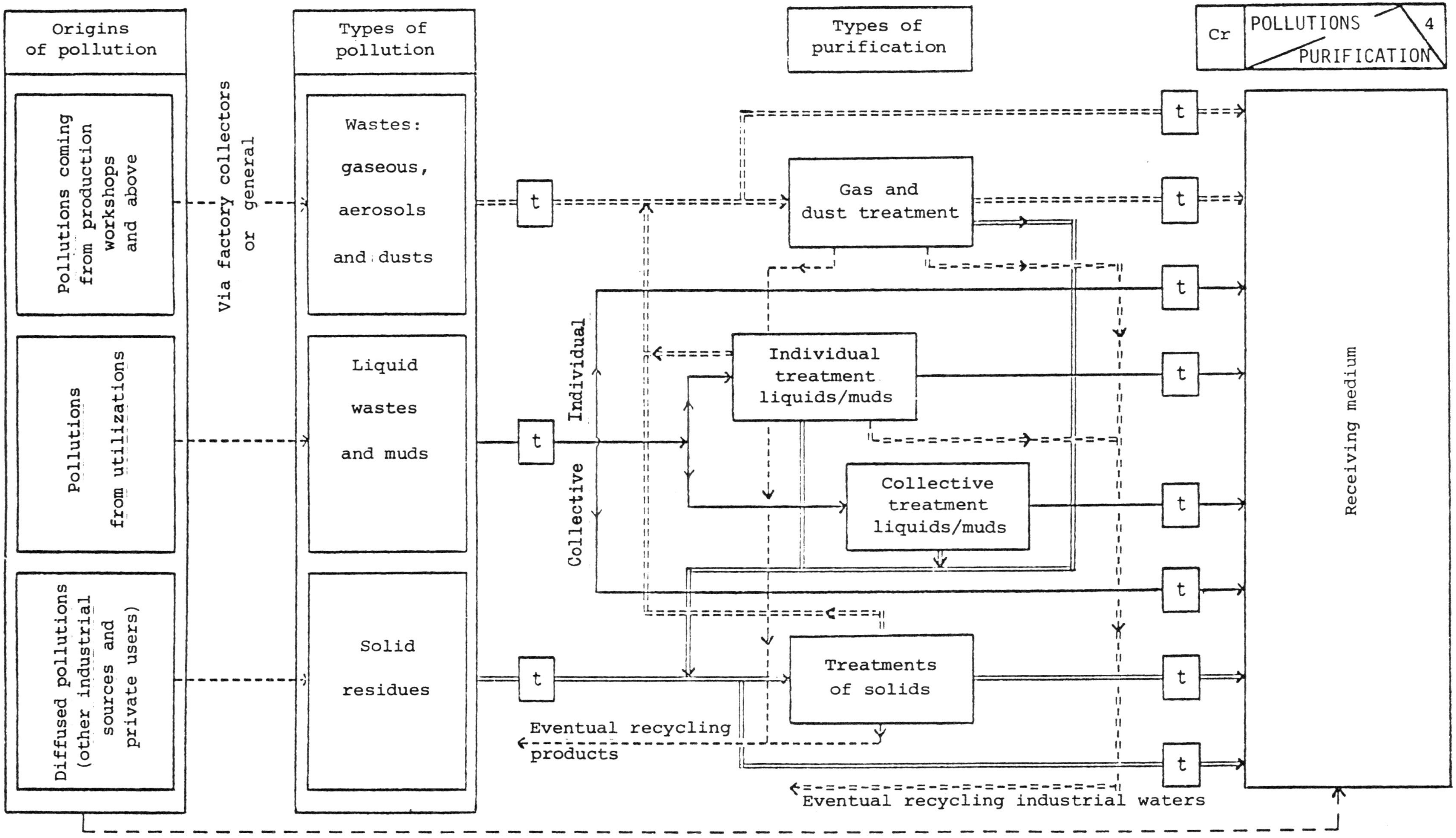
Origins of pollution
Types of pollution
Types of purification
Cr
POLLUTIONS
PURIFICATION
4
Pollutions coming from production workshops and above
Pollutions from utilizations
Diffused pollutions (other industrial sources and private users)
Via factory collectors or general
Wastes: gaseous, aerosols and dusts
Liquid wastes and muds
Solid residues
t
Individual
Collective
Gas and dust treatment
Individual treatment liquids/muds
Collective treatment liquids/muds
Treatments of solids
Eventual recycling products
Eventual recycling industrial waters
Receiving medium

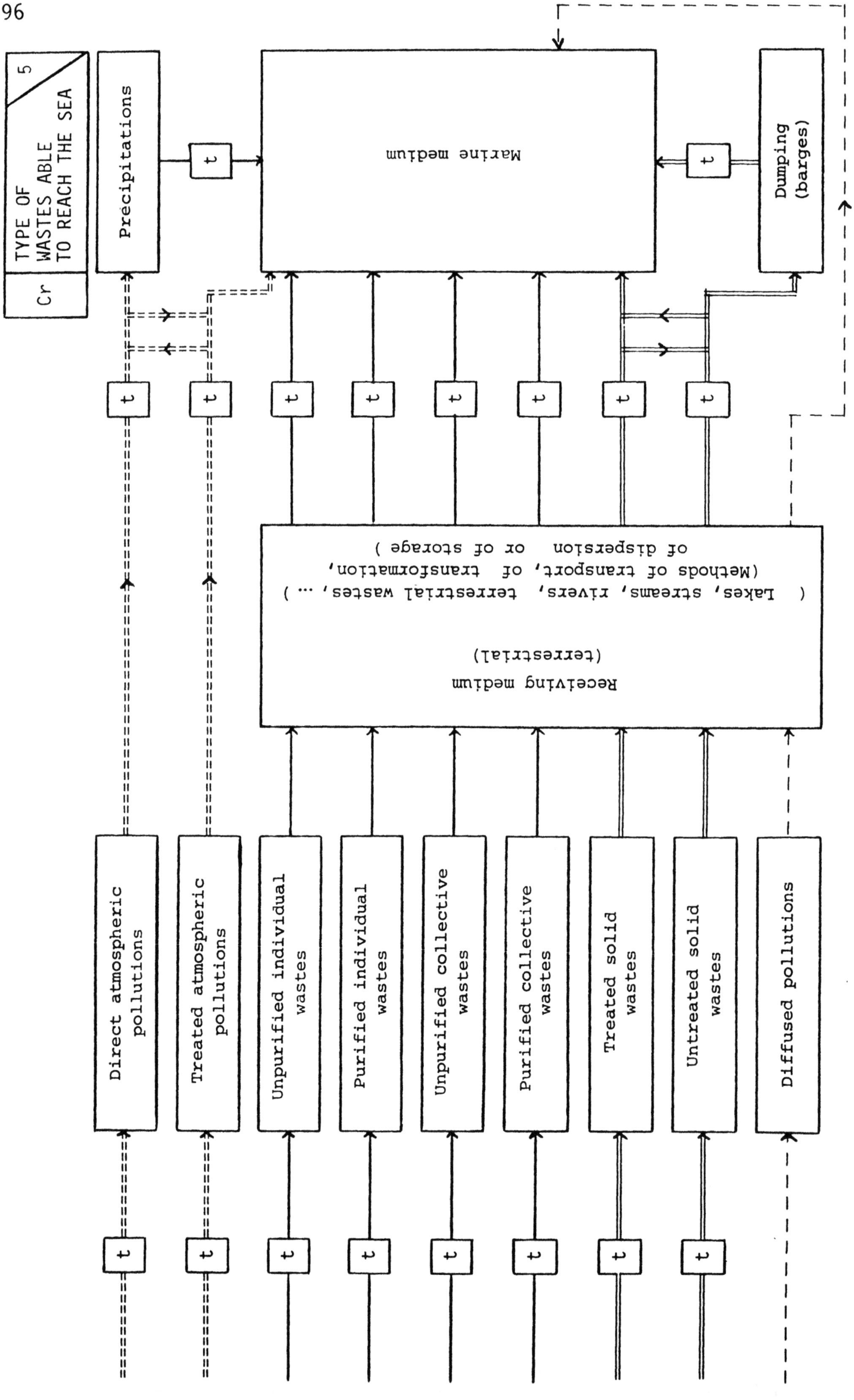
Cr
TYPE OF WASTES ABLE TO REACH THE SEA
5
Precipitations
Marine medium
Dumping (barges)
Receiving medium (terrestrial)
(Lakes, streams, rivers, terrestrial wastes, ...)
(Methods of transport, of transformation, of dispersion or of storage)
Direct atmospheric pollutions
Treated atmospheric pollutions
Unpurified individual wastes
Purified individual wastes
Unpurified collective wastes
Purified collective wastes
Treated solid wastes
Untreated solid wastes
Diffused pollutions
t

A N N E X E 2

GENERAL SYNOPTIC DIAGRAM OF POLLUTION
CAUSED BY THE PRODUCTION AND THE VARIOUS USES
OF CADMIUM AND ZINC

GENERAL SYNOPTIC DIAGRAM OF POLLUTION CAUSED BY THE PRODUCTION AND VARIOUS USES OF CADMIUM AND ZINC

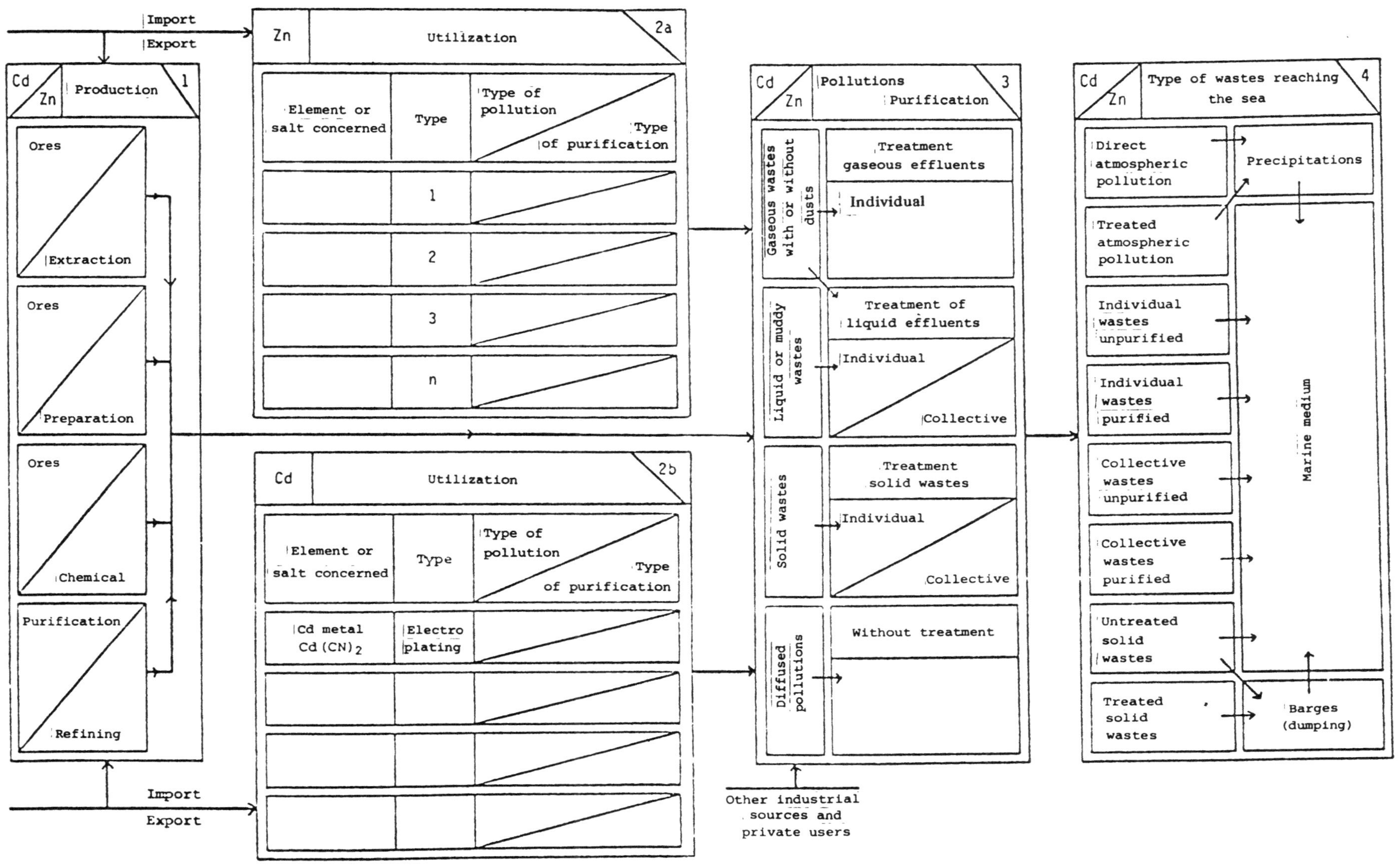

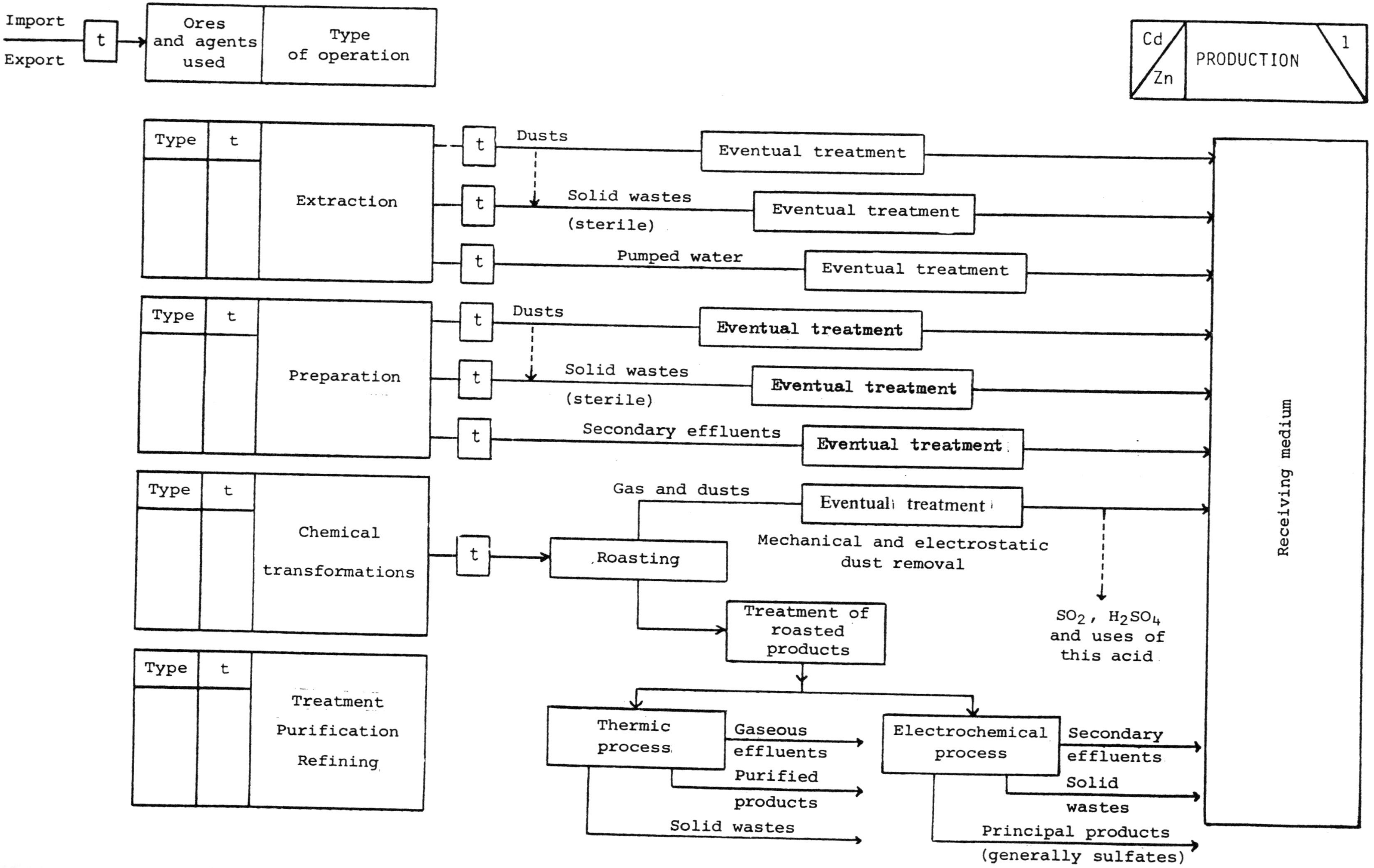
Cd
Zn
PRODUCTION
1
Import
Export
t
Ores and agents used
Type of operation
Type
t
Extraction
Preparation
Chemical transformations
Treatment Purification Refining
Dusts
Solid wastes (sterile)
Pumped water
Secondary effluents
Gas and dusts
Eventual treatment
Mechanical and electrostatic dust removal
SO_2, H_2SO_4 and uses of this acid
Roasting
Treatment of roasted products
Thermic process
Gaseous effluents
Purified products
Solid wastes
Electrochemical process
Secondary effluents
Solid wastes
Principal products (generally sulfates)
Receiving medium

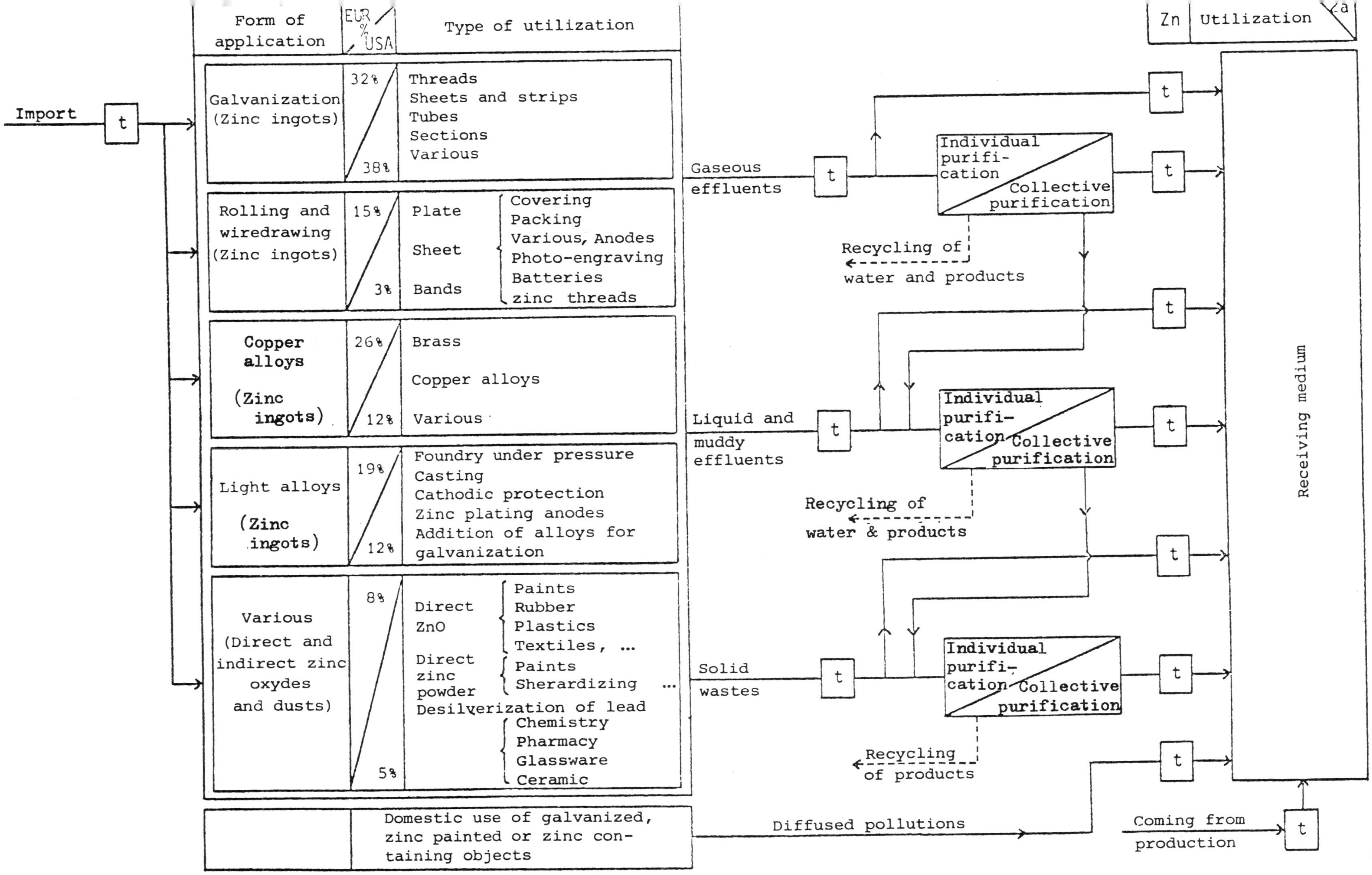

Zn
Utilization
Ia
Import
t
Form of application
EUR % USA
Type of utilization
Galvanization (Zinc ingots)
32%
38%
Threads
Sheets and strips
Tubes
Sections
Various
Rolling and wiredrawing (Zinc ingots)
15%
3%
Plate
Sheet
Bands
Covering
Packing
Various, Anodes
Photo-engraving
Batteries
zinc threads
Copper alloys (Zinc ingots)
26%
12%
Brass
Copper alloys
Various
Light alloys (Zinc ingots)
19%
12%
Foundry under pressure
Casting
Cathodic protection
Zinc plating anodes
Addition of alloys for galvanization
Various (Direct and indirect zinc oxydes and dusts)
8%
5%
Direct ZnO
Paints
Rubber
Plastics
Textiles, ...
Direct zinc powder
Paints
Sherardizing ...
Desilverization of lead
Chemistry
Pharmacy
Glassware
Ceramic
Domestic use of galvanized, zinc painted or zinc containing objects
Gaseous effluents
Liquid and muddy effluents
Solid wastes
Diffused pollutions
Individual purification
Collective purification
Individual purification
Collective purification
Individual purification
Collective purification
Recycling of water and products
Recycling of water & products
Recycling of products
Receiving medium
Coming from production

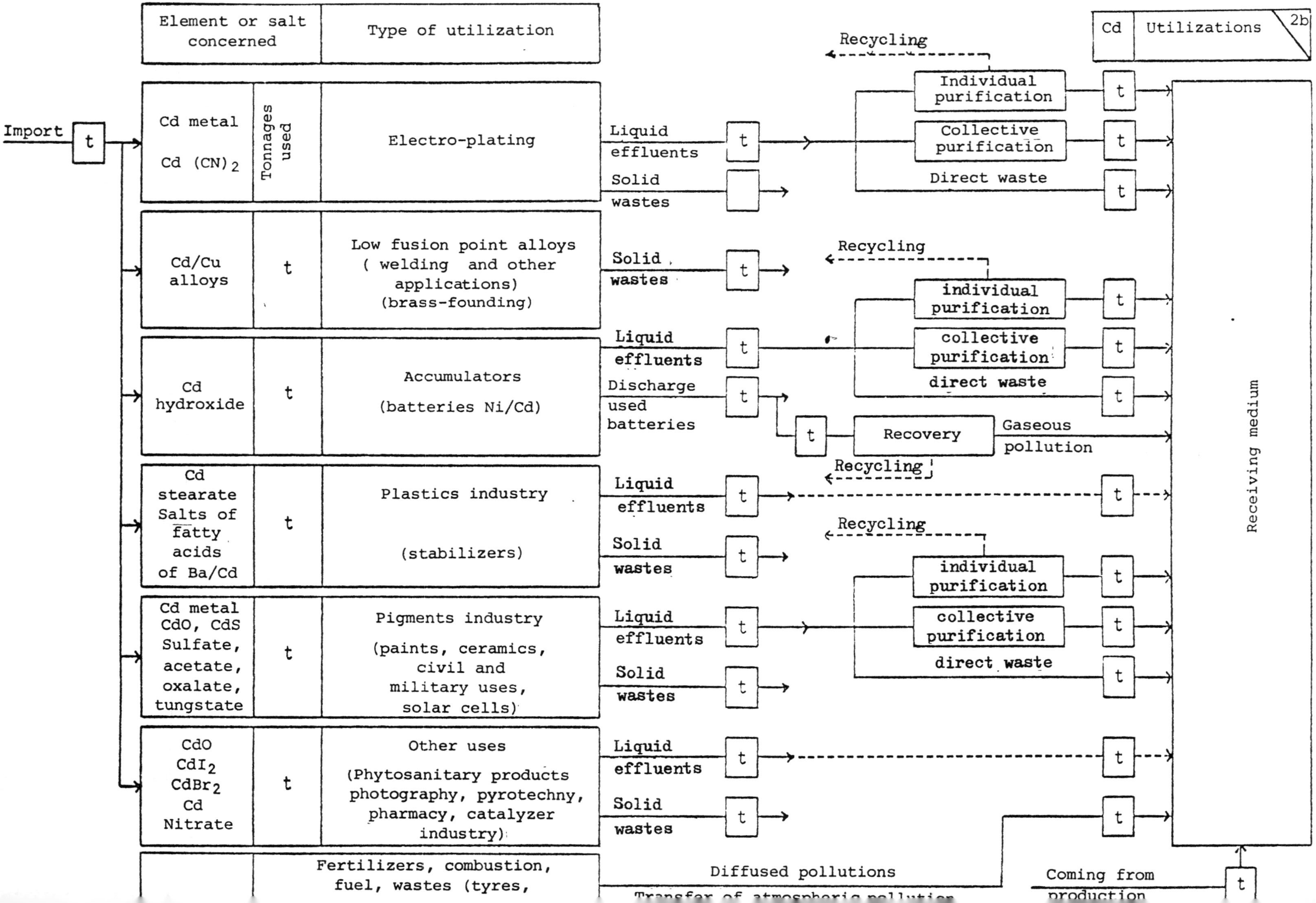
Cd
Utilizations
2b
Element or salt concerned
Tonnages used
Type of utilization
Import
t
Cd metal
Cd (CN)2
Electro-plating
Cd/Cu alloys
Low fusion point alloys (welding and other applications) (brass-founding)
Cd hydroxide
Accumulators (batteries Ni/Cd)
Cd stearate Salts of fatty acids of Ba/Cd
Plastics industry (stabilizers)
Cd metal CdO, CdS Sulfate, acetate, oxalate, tungstate
Pigments industry (paints, ceramics, civil and military uses, solar cells)
CdO CdI2 CdBr2 Cd Nitrate
Other uses (Phytosanitary products photography, pyrotechny, pharmacy, catalyzer industry)
Fertilizers, combustion, fuel, wastes (tyres,
Liquid effluents
Solid wastes
Discharge used batteries
Recycling
Individual purification
Collective purification
Direct waste
individual purification
collective purification
direct waste
Recovery
Gaseous pollution
Diffused pollutions
Coming from
Receiving medium

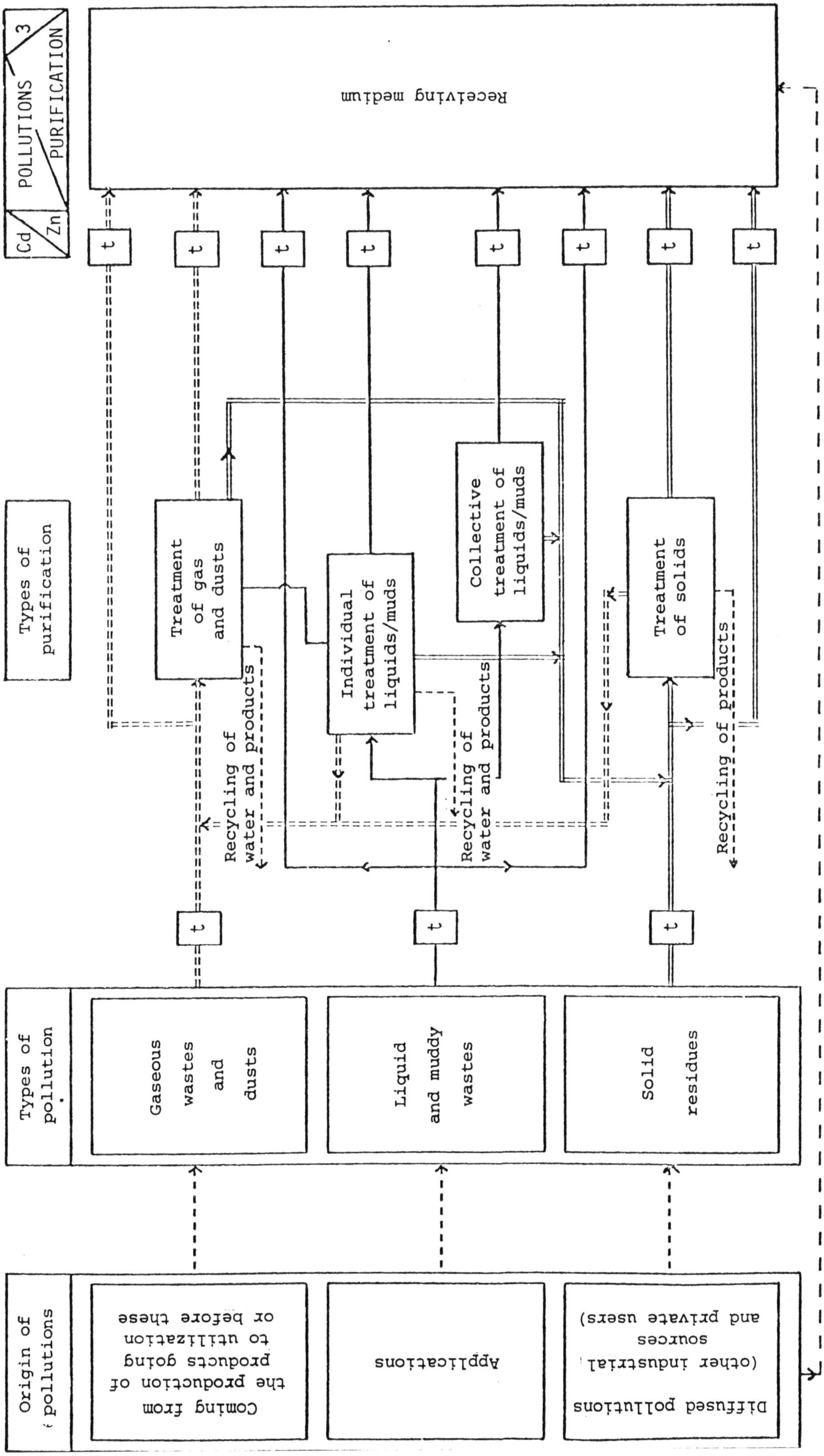
Cd
Zn
POLLUTIONS
PURIFICATION
3
Receiving medium
Types of purification
Treatment of gas and dusts
Individual treatment of liquids/muds
Collective treatment of liquids/muds
Treatment of solids
Recycling of water and products
Recycling of water and products
Recycling of products
t
Types of pollution
Gaseous wastes and dusts
Liquid and muddy wastes
Solid residues
Origin of pollutions
Coming from the production of products going to utilization or before these
Applications
Diffused pollutions (other industrial sources and private users)

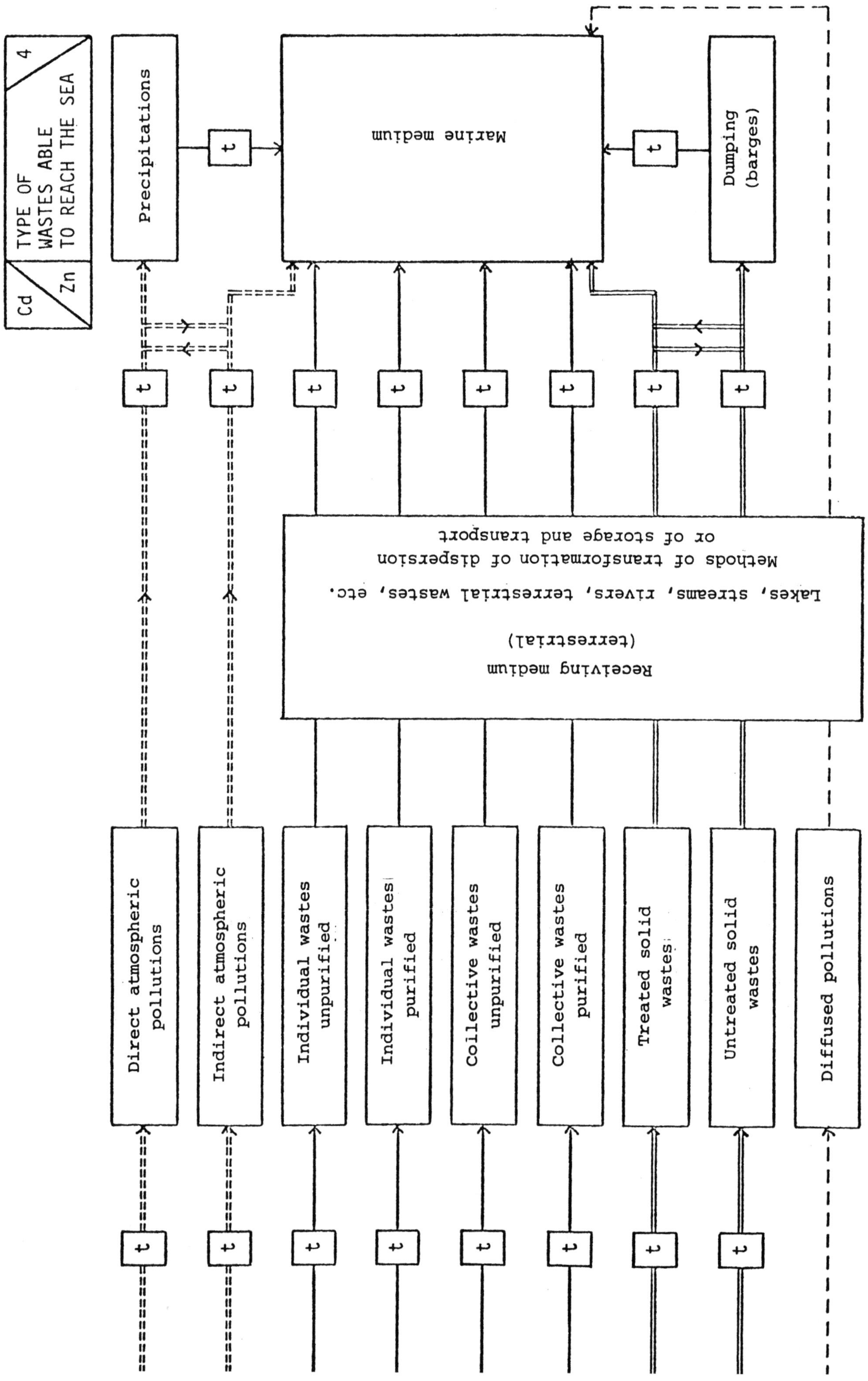
Cd
Zn
TYPE OF WASTES ABLE TO REACH THE SEA
4
Precipitations
Marine medium
Dumping (barges)
t
Receiving medium
(terrestrial)
Lakes, streams, rivers, terrestrial wastes, etc.
Methods of transformation of dispersion
or of storage and transport
Direct atmospheric pollutions
Indirect atmospheric pollutions
Individual wastes unpurified
Individual wastes purified
Collective wastes unpurified
Collective wastes purified
Treated solid wastes
Untreated solid wastes
Diffused pollutions

B I B L I O G R A P H Y

1. ASHLEY (Burton E.), 1971.- Cadmium. In : *Minerals Yearbook 1971, vol. 1 : metals, minerals and fuels*. pp. 237-243. Washington, D.C., Bureau of Mines.

2. *Commerce extérieur - tableaux analytiques. Vol. C : Produits des industries chimiques et des industries connexes*.- Luxembourg, Bruxelles, Office statistique des Communautés européennes, 1972.- (389 p.)

3. McMAHON (Albert D.), 1971.- Zinc. In : *Minerals Yearbook 1971, vol. 1 : metals, minerals and fuels*.- pp. 1239-1269.- Washington, D.C., Bureau of Mines.

4. *Metal Statistics 1962-1972*. Frankfurt / Main, Metallgesellschaft ?.- (Zinc : pp. 33-38, 227-243) - (Cadmium : pp. 45-46)

5. MORNING (John L.), 1971.- Chromium. In : *Minerals Yearbook 1971, vol. 1 : metals, minerals and fuels*.- (pp. 291-301).-

6. *Statistical summary of the mineral industry, world production, exports and imports, 1967-1971 :* cadmium.- London, Her Majesty's Stationery Office, 1973.- pp. 64-66.- (Institute of geological sciences, Mineral resources division).

7. *Statistical summary of the mineral industry, world production, exports and imports, 1967-1971 :* chrome ore and chromium.- London, Her Majesty's Stationery Office, 1973.- pp. 73-77.- (Institute of geological sciences, Mineral resources division).

8. *Statistical summary of the mineral industry, world production, exports and imports, 1967-1971 :* zinc.- London, Her Majesty's Stationery Office, 1973.-pp. 384-397.- (Institute of geological sciences, Mineral resources division).

9. Statistiques - Statistical tables (Paper prepared on the occasion of) *1er symposium mondial : Energie et matières premières*, Paris, 6-8 June 1974. - (20 p.)

10. Zinc. In : *Yearbook of the American bureau of metal statistics*.- pp. 69-94.-New York, American bureau of metals statistics, 1972.

11. *Metal bulletin handbook, 1973*, 6th ed.- London, Metal bulletin Ltd., 1973.- 970 p.

12. *Chromium minerals, ferro-chrome, chromium and chromium chemicals : world survey of production and consumption with special reference to future demand and prices*. - London, Roskill Information Services Ltd., 1972.- viii-236 p.

13. PATTERSON (J.W.) & MINEAR (R.A.), 1973 - Treatment technology for hexavalent chromium. In : *Wastewater treatment technology*, pp. 51-68, bibliogr.- Chicago, Ill., Illinois Institute for environmental quality.

14. PATTERSON (J.W.) & MINEAR (R.A.), 1973.- Treatment technology for trivalent chromium. In : *Wastewater treatment technology*, pp. 69-83, Bibliogr.-Chicago, Ill., Illinois Institute for environmental quality.

15. RYDBERG (J.), REINHARDT (H.), LUNDEN (B.) & al., 1973.- Recovery of metals and acids from stainless steel pickling bath. In : *International symposium on hydrometallurgy*, Chicago, Ill., February 25-March 1, 1973, pp. 589-601.- New York., the American Institute of mining, metallurgical & petroleum engineers Inc.

16. *Rapport sur le contrôle du chrome dans les effluents d'usines de revêtements électrolytiques aux U.S.A.*, prepared by "The American Electroplaters' Society, environmental pollution central subcommittee". (Translation of an article which appeared in *Plating*, Sept. 1970, pp. 889-892) - 9p., ill. bibliogr.

17. *The economics of cadmium.* And ed.-London, Roskill Information Services Ltd., 1974.- viii - 209 p.

18. Le marché du zinc en France en 1972. *Zinc information, Généralités*, n° 3, (1973), 12 p.

19. MONTAGUE (H.L.), 1971 (?) - *The extractive metallurgy of zinc : review of processes and projections for the future.*- New York, N.Y., the Metallurgical society of AIME. 52 p. - (Paper n° A71-74)

20. LUND (Herbert F.), Ed., 1971.- *Industrial pollution control handbook.*- New York, McGraw-Hill Book Co.- xxic-862 p., ill.

21. RICE (N.M.) & BROADFOOT (A.), 1972.-Hydrometallurgical recovery of zinc from blast furnace slags. Paper presented at the *1972 Conference of metallurgists of the C.I.M.*, Halifax, Nova Scotia, August, 1972.- 36 p.- (University of Leeds, Department of mining & mineral sciences).

22. *Le Zinc, sa métallurgie, ses applications.*- Levallois, Centre technique du zinc, 1971.- 72 p., ill.

23. *Le zincage électrolytique.*- Levallois, Centre technique du zinc, s.d.- 15 p.

24. BLOK (J.), 1973.- Inhibitie van actief slib systemen door cyanaat en cyanide. (Toxicity of CN and CNO for activated sludge). *H_2O* vol. 6, n° 20, pp. 524-526.

25. *Destruction des cyanures.*- Paris, Produits chimiques Ugine Kuhlmann, s.d. - 24 p.

26. Destruction de HCN par combustion catalytique. *Informations chimie*, n° 125, novembre 1973, p. 135.

27. GERMAIN (J.E.), 1973.- Plating and cyanide wastes. *Journal Water pollution control federation*, vol. 45, n° 6, pp. 1192-1198, bibliogr.

28. *Impianti di trattamento eseguiti nelle FIAT - Sezione Officine di Bari.*- S.1, Fiat, Dipartimento ecologia s.d.- 7p., 4 fig.

29. *Impianto di trattamento delle scorie e delle acque cianurate.*- S.1, fiat, Dipartimento ecologia, s.d. - 3 p., 4 fig.

30. LECLERC (G.), 1971.- Les effluents de traitements de surface : problèmes et résolutions. *Galvano,* n° 407, pp. 39-46, fig.

31. MARINE (C.M.), LECHI (G.M.) & FOX (R.), 1972.- Trattamento delle acque di scarico degli stabilimenti Fiat localizzati sulla costa adriatica e proposta di un metodo di prospezione aerea per studiare l'inquinamento del Mare Adriatico. (Paper presented at : *Concegno parlamentare italo-jugoslavo sull'inquinamento del Mare Adriatico, 22-24 ottobre 1972*). 14 p., 4 fig.

32. PATTERSON (J.W.) & MINEAR (R.A.), 1973.- Treatment technology for cyanide. In : *Wastewater treatment technology,* And ed., pp. 107-130, bibliogr.-Chicago, Ill., Illinois Institute for environmental quality.

33. ZUMBRUNN (J.P.), 1973.- Nouvelles applications des produits peroxydés pour l'épuration des eaux résiduaires. *Informations chimie,* n° 124, pp. 189-194.

34. AMIEL (J.), 1959.- Propriétés physiologiques du chrome. In : *Nouveau traité de chimie minérale,* ed. by P. Pascal, pp. 94-99, bibliogr.-Paris, Masson & Cie.

35. ANSELL (M.) & LEWIS (F.A.S.), 1970.- A review of cyanide concentrations found in human organs, a survey of literature concerning cyanide metabolism. "normal", non-fatal, and fatal body cyanide levels. *Journal of Forensic Medicine,* vol. 17, n° 4, pp. 148-155.

36. BROWNING (E.), 1969.- Chromium. In : *Toxicity of industrial metals,* 2nd ed., pp. 101-112, bibliogr.-London, Butterworths.

37. L'emploi des cyanures dans l'industrie (35-03). *Cahiers de notes documentaires,* Institut national de sécurité, note n° 462-41-65, n° 41, 1965, pp. 247-252.

38. FRIBERG (Lars), 1973.- *Cadmium in the environment, II.*- Washington, D.C., U.S. Environmental protection agency, Office of research and monitoring. vi-169 p., ill.

39. KAZANTZIS (G.), 1973.- *Chrome and nickel.*- S.1, Laboratoire Courrières Kuhlmann, pp. 2451-2457.- (Mémoire n° 443) -

40. MASCARENHAS (Bento R.), GELLER (Alan C.) & GOODMAN (Alvin I.), 1969.- Cyanide poisoning, medical emergency. *New York State Journal of medicine,* June 15 1969, pp. 1782-1784.

41. SAIA (B.) & DE ROSA (E.),1970.- Remarques sur l'intoxication chronique par les cyanures. (Translation of an article published in : *Medicina del lavoro,* vol. 61, n° 11, pp. 580-586).

42. BLOK (J.), 1974.- Respirometric measurements on activated sludge. *Water research,* vol. 8, pp. 11-18.

43. L'élimination des déchets industriels. *Informations chimie,* n° 124, 1973, pp. 167-172.

44. *Encyclopédie internationale des sciences et des techniques.*- Paris, Presses de la Cité, 1969-1974.- 10 vol + index.

45. EYE (J. David), 1973.- Tannery wastes. *Journal Water Pollution Control Federation,* vol. 45, n° 6, pp. 1173-1176.

46. MACAULEY (Dennis C.), 1973.- Chemicals and allied products. *Journal Water Pollution Control Federation,* vol. 45, n° 6, pp. 1218-1230, bibliogr.

47. *Mining annual review, 1973.*- London, The Mining Journal Ltd., 1973.- 498 p.- (Supplement to *Mining Journal*).

48. PATTERSON (J.W.) & CHENG (M.H.), 1973.- Steel industry wastes. *Journal Water Pollution Control Federation,* vol. 45, n° 6, pp. 1184-1188.

49. PATTERSON (J.W.) & MINEAR (R.A.), 1973.- *Wastewater treatment technology,* 2nd ed.- Chicago, Ill., Illinois Institute for environmental quality. iv 222 p.

50. REISH (Donald J.), 1973.- Marine and estuarine pollution. *Journal Water Pollution Control Federation,* vol. 45 n° 6, pp. 1310-1319, bibliogr.

51. ROCK (David M.), 1970.- Hydroxide precipitation and recovery of certain metallic ions from wastewaters. *Chemical engineering progress symposium series, Water,* vol. 67, n° 107, pp. 442-444.

52. VAN STONE (G.R.), 1972.- Treatment of coke plant waste effluent. *Iron & steel engineer,* vol. 49 n° 4, pp. 63-66.

53. *Rapport au sujet d'une étude sur les sources et les volumes de résidus et déchets solides, liquides et gazeux de métaux lourds rejetés dans le milieu ambiant en République fédérale d'Allemagne et en France, réalisé pour le compte de la Commission des Communautés européennes.*- Bruxelles, Gummersbach, Kienbaum, conseiller d'entreprises, 1972.- 72 p.- (Report III/1898/72-F).

54. WIENTJENS (S.H.J.M.), 1972.- *Le mercure, le cadmium et le chrome aux Pays-Bas. Final Report.*- Delft Centraal Laboratorium T.N.O.- 33 p.

55. Vatten-och luftföroreningar från ytbehandlingsanläggningar. *Statens Naturvårdsverk, Publikationer,* 1972, n° 4. 142 p., fig.

56. CHANG REYES (L.), 1972.- Contaminantes potenciales que pueden afectar a los organismos del ambiente marino a lo largo de la costa del Peru. In : *Marine Pollution and sea life,* pp. 106-111. - London, Fishing News (Books) Ltd.

57. *Annuaire - Minerais et métaux.*- Paris, Penarroya, 1971.- 227 p.

58. *Problèmes posés par la contamination de l'homme et de son milieu par le mercure et le cadmium.* Mémoires. Luxembourg, 3-5 July 1973.- Luxembourg, Commission des Communautés européennes.

59. *Le mercure et l'environment. Etudes sur l'utilisation du mercure, ses émissions, ses effets biologiques et son contrôle.*- Paris, Organisation de coopération et de véveloppement économiques, 1974.- 211 p., bibliogr.

60. *Les problèmes sanitaires posés par le plomb présent dans l'environnement.* Actes. Symposium international. Amsterdam, October 2-6, 1972 xiv-1168 p.- (organised jointly by the Commission of the European Communities and United States Environmental Protection Agency).

61. AUBERT (M.), BITTEL (R.), LAUMOND (F.) & al., 1974.- Utilisation d'une chaîne trophodynamique de type néritique à mollusque pour l'étude des transferts des polluants métalliques. *Revue internationale d'océanographie médicale,* vol. 33, pp. 7-29, bibliogr. (Compte-rendu du VIe colloque international d'oceanographie médicale, 26-30 septembre 1973, Portoroz (Yugoslavia).

62. BREMOND (R.) & VUICHARD (R.), 1973.- *Paramètres de la qualité des eaux.*- Paris, La Documentation française.- 180 p., ill.- (Environment) - (Ministère de la Protection de la nature ét de l'environnement, Secrétariat permanent pour l'étude des problèmes de l'eau).

63. GESAMP, 1972.- *Report of the fourth session,* held at ... Geneva, 18-23 September 1972.- 80 p.- (Joint Group of experts on the scientific aspects of marine pollution).

64. IVANOFF (Alexandre), 1972.- *Introduction à l'océanographie : propriétés physiques et chimiques des mers.* Vol. 1.- Paris, Vuivert.- 206 p.

65. HALSTEAD (B.N.), 1972.- Toxicity of marine organisms caused by pollutants. In : *Marine pollution and sea life,* pp. 584-594, bibliogr.- London, Fishing News (Books) Ltd.

66. LAUWERYS (R.), 1972.- *Toxicologie industrielle et intoxications professionnelles.*- Jembloux, Duculot.- 609 p.

67. PATTY (F.), 1958.- *Industrial hygiene and toxicology.*- New York, London, Interscience.- 2 vol.

68. WALDICHUK (Michael), 1974.- Coastal marine pollution and fish. *Ocean Management,* vol. 2, n° 1, 1-60, bibliogr.

69. TODD (David Keith), Ed., 1970.- *The Water Encyclopedia.*- Port Washington, N.Y. Water information center.- p. 62.

70. idem, p. 185.

71. Information provided by The State Institute for drinking water research, The Hague, and The State Institute for public health, Bilthoven, The Netherlands.

72. CIACCIO (Leonard L.), Ed., 1971.- *Water and water pollution.* New York, Marcel Dekker Inc.- Vol. 1.

73. CARPENTIER (James H.), 1972.- *National Bureau of Standards special publication 351,* chapter 7.

74. HOOGEVEEN (L.P.J.) & BUTZELAAR (P.F.), 1974.- *Philips Technical Review 34-123.*

75. Instruments perfected by The Axel Johnson Institute, Nynasham, Sweden.

76. KROON (D.J.), 1974.- *Philips Technical Review 34-113.*

LIST OF INDUSTRIAL GROUPS MEMBERS OF THE A.E.O. AND PERSONS WHO CONTRIBUTED TO THE STUDY

LIST OF INDUSTRIAL GROUPS MEMBERS OF THE A.E.O. AND PERSONS WHO CONTRIBUTED TO THE STUDY

EUROPEAN OCEANIC ASSOCIATION

M. Jacques CONSTANS *Deputy Director General of the A.E.O.*	A.E.O. Secretariat

CONSULTATIVE COMMISSION AND READING COMMITTEE OF THE A.E.O.

Rear-Admiral Wijnand LANGERAAR *Permanent adviser to the A.E.O.* *Former chief hydrographer of the Royal Dutch Navy*	President of the consultative commission
M. Jean BENOIT	Scientific and Technical Secretary
Dr. Claude MILLET	Expert on marine pharma-toxicology

INDUSTRIAL GROUPS

AKZO N.V. *(Netherlands)*	Ir. E.L.J. BANCKEN *Research coordinator*
COMPAGNIE GENERALE D'ELECTRICITE *(France)*	M. F. CARRIVE *Director to the C.G.A.*
FIAT S.p.A. *(Italy)*	M. R. FOX *Deputy Director* *Department of Ecology* *Siteco S.p.A.*
PECHINEY UGINE KUHLMANN *(France)*	M. A. NOE *Director of the Nuclear and New Techniques Branch*
PHILIPS *(The Netherlands)*	Dr. D.J. KROON *Department of environmental Sciences* *Philips Research Laboratories*

A. JOHNSON & CO. *(Sweden)* SKANDINAVISKA ENSKILDA BANKEN *(Sweden)*	Dr. G. OLGÅRD *Incentive A.B.*
RIJN SCHELDE VEROLME *(Netherlands)*	Mr. Theo OOSTERS *Delta Engineering B.V.*
TECHNITAL S.p.A. *(Italy)*	Dr. R. RISMONDO *Director of Applied Ecology Department* Dr. A. TROMBETTA *Consulting Engineer*
TECNOMARE S.p.A. *(Italy)*	Dr. R. PESARESI *Marine Systems*
Professor Jean ROUVIER *Faculty of Law & Economic Sciences, Paris* *International legal consultant*	Legal adviser of the A.E.O.
Professor Henri SCHWAMM *Graduate Institute of International Studies, University of Geneva* *International economic consultant*	Economic adviser of the A.E.O.

OTHER PARTICIPANTS

Mr. Pierre BONNET and Mr. Jean VACCAREZZA
Officials of the Environment and Consumer Protection Service of the Commission of the European Communities

GPSR Compliance

The European Union's (EU) General Product Safety Regulation (GPSR) is a set of rules that requires consumer products to be safe and our obligations to ensure this.

If you have any concerns about our products, you can contact us on ProductSafety@springernature.com

In case Publisher is established outside the EU, the EU authorized representative is:

Springer Nature Customer Service Center GmbH
Europaplatz 3
69115 Heidelberg, Germany

Batch number: 10615097

Printed by Printforce, the Netherlands